Agriculture in Australia

Agriculture in Australia

Bill Malcolm, Peter Sale, Brian Leury & Snow Barlow

OXFORD
UNIVERSITY PRESS
AUSTRALIA & NEW ZEALAND

OXFORD
UNIVERSITY PRESS
AUSTRALIA & NEW ZEALAND

253 Normanby Road, South Melbourne, Victoria 3205, Australia

Oxford University Press is a department of the University of Oxford.
It furthers the University's objective of excellence in research,
scholarship, and education by publishing worldwide in

Oxford New York

Auckland Cape Town Dar es Salaam Hong Kong Karachi
Kuala Lumpur Madrid Melbourne Mexico City Nairobi
New Delhi Shanghai Taipei Toronto

With offices in

Argentina Austria Brazil Chile Czech Republic France Greece
Guatemala Hungary Italy Japan Poland Portugal Singapore
South Korea Switzerland Thailand Turkey Ukraine Vietnam

OXFORD is a trademark of Oxford University Press
in the UK and in certain other countries

First published 1996
Reprinted 2000, 2002, 2003, 2004, 2005
Second edition published 2009
Reprinted 2013(D)

National Library of Australia Cataloguing-in-Publication entry

Agriculture in Australia : an introduction / Bill Malcolm ... [et al.].
2nd ed.
9780195560114 (pbk.)
Includes index.
Bibliography.
Farm management–Australia.
Agriculture–Australia.
Malcolm, L. R.
630.94

Edited by Tim Fullerton
Text design by Patrick Cannon
Typeset by Cannon Typesetting
Proofread by James Anderson
Indexed by Russell Brooks
Printed and bound in Australia By Ligare Book Printers, Pty td.

Contents

List of Figures

List of Tables

Preface

The motivation for writing this book is the belief that the major resource that could potentially constrain agricultural production in the future is the quantity and quality of human capital working in the agricultural sector. The challenge of feeding the future world population, and the challenging nature of agriculture, means that there is need for the best and brightest young minds to work in the science and economics of agriculture in the future.

Farming successfully in Australia is a challenge for farmers, scientists and economists alike. We try in this book to provide a comprehensive, integrated view of the complex and challenging task of farming in Australia. Farmers have to try to make a living today, and tomorrow. The major challenges to farmers come from volatility in natural conditions and markets from season to season. However, good management of adequate capital succeeds and so the agricultural sector remains extremely important for the economic well-being of the nation. Family farming business-people who can master the information and the technology sufficiently well to increase productivity consistently, and who do not have excessive debt, have a rewarding way to live and to make a living. As well, business-people involved in providing the services needed in agriculture to produce farm products and to transform them to meet consumer requirements benefit from efficient agricultural processes, as do consumers. Interesting, challenging and rewarding careers are to be had working in the science, economics and management of agricultural production and marketing processes. Our hope is that the analyses and explanations about agriculture in this book will sow the seeds of curiosity and form the basis of an interesting and valuable learning experience.

About this book

The emphasis in this book is on more than describing 'what is': instead the emphasis is on analysing and explaining agriculture in Australia in accessible ways befitting an introduction to an area of study. The book comprises six

Parts, covering an introduction, resources and farming in Australia, plants, animals, whole farm, and change. In Part 1, Introduction, we start with the big challenges facing farmers in Australia and around the globe: feeding the growing population of the world and adapting to changes in natural conditions. In Part 2, Resources and Farming in Australia, we look at agriculture in terms of 'what natural resources we have' and 'what happens where and why'. The natural resources, climatic conditions, the economic system, location, the effects of agriculture on growth in the rest of the economy, and the effects of growth in the economy on agriculture are all important 'rules' of the farming game.

In Part 3, Fundamental Processes of Plant Production, we explain how plants live and grow. The roles of the major species of field crops and pastures are outlined. In Part 4, Fundamental Processes of Animal Production, we explain how animals live and grow, and how the major animal products, meat, wool, and milk, are produced.

In Part 5, The Farm Whole Business, the economic concepts that are critical to understanding profitable production processes are explained, and their applications in practice are demonstrated. The focus is on integrating the technical, economic and human components of a farm business into whole farm analysis, and applying the whole farm approach to analysing the state of a farm business and to decisions about using farm resources profitably.

Part 6, Changing Agriculture, is concerned with sustaining and improving cropping and animal activities. A further concern is coping with a changing climate, to the extent that agriculture can continue to contribute to the well-being of people in Australia and the rest of the world. A central theme is that sustaining the quality of resources used in agriculture and increasing the productivity of agriculture are compatible, and complementary. Sustaining resources is about how to maintain the productivity of resources by protecting them from degradation. Improving agriculture is about how to sustain the major agricultural activities by improving productivity and profitability. In this regard we focus on extensive cropping, and extensive and intensive pasture production activities. Coping with climate change is about the potential implications of farmers having to farm in the future under hotter, drier and more volatile natural conditions.

In this book agriculture is viewed as a combination of production, profit and sustainability—three concepts that are not distinct, nor mutually exclusive, but are directly and closely related. Agricultural businesses will not be sustainable over the long haul unless they are productive and profitable, and agricultural businesses will not remain productive and profitable over time unless they use resources in sustainable ways. Established agricultural scientific and economic knowledge needs to be applied to managing resources in mainstream agricultural systems. To this end we focus on commercial farming operations, where the major proportion of total income comes from the farm. Most farms are family owned, though integrated farm businesses are important in the intensive industries.

In particular, we look at those farming operations that contribute a significant share of the nation's gross value of agricultural production. That is, extensive cropping; sheep and cattle grazing; intensive irrigated crops and pasture; and intensive dairy, pig and poultry activities. Horticultural activities are not specifically explored in the same way as are extensive cropping activities and extensive and intensive pasture production activities, although the principles of plant production, land and water management, and business management and marketing, which are explained in depth, apply equally to all forms of production.

Alternative production systems, such as organic or biodynamic production systems, exotic industries, and minor industries, are not examined—important as these are to their small bands of enthusiasts. Further, there is greater emphasis on the extensive cropping, pasture and animal production systems, and on the intensive production systems, of southern Australia and the coastal regions of the continent, than on the extensive pastoral enterprises that occur across the arid northern and inland rangelands.

The perspectives brought to bear on agriculture in this book are primarily those of the disciplines of animal and plant science, agricultural economics, and business management and marketing. At times the emphasis is on agricultural science and the technical aspects of solving problems in production; at times the emphasis is on agricultural economics, business management and marketing. Throughout the book, the whole farm business, and beyond it, are kept in mind. In our understanding of the whole farm business approach it is better to solve the whole problem in an approximate way than to solve part of the problem in a precise way.

In *Agriculture in Australia* gender roles in agriculture are not specially mentioned, nor the food supply activities of the Indigenous Australians before European settlement and their participation subsequently in the agricultural production processes. This book is not about the sociology of rural life—such investigation has different emphases to those in this book—though, the importance of the goals, interests, skills, attitudes to risk, and stage of life of the farm family in determining productivity is emphasised repeatedly. Our focus is very close to the farm family and the family farm, as being extremely durable business organisations operating under difficult environmental and economic conditions, where the first measure of success is the continuation of the family business for as long as a farm family wants.

PART 1

Introduction

Understanding Agriculture

1

CONTENTS

Introduction

Agriculture in Australia is the business of producing food and fibre in a battle against the vagaries of climate and markets. Agriculture the Australian way is agriculture the hard way. In Australia, the old continent, the soils are not naturally fertile. In Australia, the dry continent, rainfall is low, volatile and uncertain. In Australia, the big, distant country with the small population, most farmers have to sell most of their production to the rest of the world, far away. Agriculture the Australian way is about farmers continually confronting difficult and constantly changing natural, economic, technical and social circumstances. Agriculture in Australia is a monument to a strong tradition of scientific research producing a steady stream of productive new agricultural ideas and technologies. It is a story of farmers continually increasing their productivity more rapidly than in any other part of the economy. It is a story of producing food and fibre extremely efficiently. For farmers in Australia, and their fellow travellers, farming is their livelihood and their life. It is their passion. Most are very good at it.

Business-people and scientists working in agriculture, horticulture, and natural resource management in Australia, and in industries related to agriculture and horticulture, contribute to economic growth and improve people's lives by:

- increasing the **supply** of food and fibre. This helps to meet increases in **demand** for food and fibre from growing populations and from increasingly wealthy populations all over the world.
- expanding exports. Earning income by selling farm products to people overseas makes it possible for people in Australia to buy imported goods and services from overseas.
- making profits. This provides **capital** for further investment, which enables businesses and the economy to grow and people to become wealthier.
- providing a market for the industrial sector. Farmers buy inputs from other businesses in the economy to use in production.
- maintaining, improving and protecting the quality of land, water and native flora and fauna. Farmers manage most of Australia's land.

In the Australian economy 120,000 farmers contribute agricultural and horticultural production valued at the farm each year at around $30 billion. This is between 2 and 3 per cent of the value of total production in the Australian economy (gross domestic product). In 1980 there were 180,000 farmers producing 5 per cent of the nation's gross domestic product.

Most farm production is exported: 90 per cent of wool and cotton, 80 per cent of wheat, 50 per cent of barley and rice, 40 per cent of beef and grain legumes, 30 per cent of dairy products and 20 per cent of fruit production. Raw agricultural and horticultural products are 25 per cent of Australia's total exports. This share of exports is 35 per cent of exports when manufactured agricultural products are included. Four per cent of the working people of Australia work on farms. Beyond the farm gate are a further 10 per cent of workers in jobs related directly to agriculture and horticulture.

Farmers all over the world face two great challenges: feeding a rapidly growing population and adapting to changed climate. Scientists think that the climate challenge for farmers in Australia is that the unusual run of hot and dry and difficult farming seasons of most of the past decade may become the usual run of seasonal conditions in fifty years' time:

- By 2030 it is likely that average temperatures will be 1°C higher than in 1990, hotter in parts inland and cooler on the coast.
- The probability that temperature will be higher than 1°C by 2030 is 10–20 per cent in coastal areas and 50+ per cent inland.
- By 2070 increases in average temperature will be between 1.8°C and 3.4°C.

The extent of the population-food challenge is set out in the rest of this chapter. The climate challenge is discussed in chapter 13.

World population

Farmers in Australia continually confront change and new challenges. They adapt to change to continue to succeed. Australian farmers face two major challenges in the next forty years. One challenge will be helping their fellow farmers from all around the world to supply enough food and fibre, efficiently and cheaply enough, to feed a population of around 9.5 billion people by 2050. The other big challenge will be to increase food production while dealing with and adapting to significant changes in climate, natural resources and markets.

Ten thousand years ago there were 5 million people in the world. Up to this time people caught their food. As people improved at catching animals and ate better, population grew faster. From 5000 to 10,000 years before the present, known as the Neolithic Age, people started farming, using slash-and-burn cultivation. World population grew from 5 million to 50 million people. From 5000 to 3000 years ago, the population of the world doubled again to around 100 million people as systems of slash-and-burn cultivation developed and large societies grew based on irrigated agriculture in the valleys of the Indus, Mesopotamia and the Nile.

In the next 2000 years, from 1000 BC to AD 1000, world population more than doubled, reaching 250 million people. This growth came from people learning to grow irrigated rice in the valleys and deltas of China, India and South-East Asia, and from irrigated farming beginning in the Americas. During this period, people adopted farming systems based on rain, cultivation and fallowing in the Mediterranean and Northern Europe. These systems replaced the slash-and-burn cultivation systems.

During the Middle Ages (1100 to 1300), climate warmed and agriculture in Europe came on apace. Land was cultivated using animals to pull implements. Population more than tripled in Europe in this short time. Irrigated rice growing in Asia, and fallow and cultivation cropping systems in Europe, meant that people in these regions fed better and bred better. Consequently, by the start of the sixteenth century, world population reached 500 million people.

In the sixteenth century Europeans started to expand their territories into the temperate regions of America, South America, and then later into Australia and New Zealand. With settlement of the New World, world supply of food and fibre increased rapidly. Populations grew rapidly.

In the past several hundred years the rate of world population growth has become faster and faster, as the following figures show. In the year AD 1 world population was 250 million people. World population grew by 0.04 per cent per year to reach 500 million people by AD 1650. In less than 200 more years, the population of the world reached one billion people. It had taken from the beginning of time to 1800 for the population of the world to reach this figure. By 1930 world population was 2 billion people. At a growth rate of 2 per cent per year population doubles in 35 years. By 1975 the population was 4 billion people. World population reached 5.7 billion people by 1993 and 6.5 billion people by 2007.

The United Nations projects that world population is most likely to peak at around 9.5 billion people by 2075. It is then likely to stabilise at around this level for the next 300 years. If this is so, it means the challenge farmers face to increase food supply per capita for a growing population in the next 65 years will be the greatest challenge farmers and their supporting scientists and people in agribusiness worldwide will face over the next 300 years.

In absolute numbers, world population increased by 71 million people per year on average between 1950 and 2000, when world population grew at a rate of 1.76 per cent per year. The population of the world is likely to increase by 57 million people per year on average between 2000 and 2050. The annual rate at which the population is growing is slowing, but world population will increase by 47 per cent by 2050. The expected average growth rate to 2050 is 0.77 per cent per year.

The number of children per woman is the key to the rate that population grows. This measure will decline from 2.8 children per woman in 2000 to around two children per woman in 2050. In the longer term, fertility is expected to stabilise at 1.85 children per woman. This will stabilise world population at around 9.5 billion people by 2050. A slight increase in the stable level of fertility will mean world population will reach 10.6 billion by 2050. A less than expected fertility rate will produce a population of 7.4 billion by 2050.

The current fertility rate in the poorest countries is over three children per woman. Ninety-nine per cent of the increase in world population in the next 40 years will occur in the countries of the world that are currently the poorest. The population of Asia will increase by 1.5 billion people over the next 40 years, and the population of Africa will grow by 1 billion people in this time. Growing the food to feed (most of) this rapidly growing population will be done by farmers in rich and poor countries adopting modern scientific agriculture using machinery and technology, and specialising and trading.

Food supply and demand

Food supply

World population growth has major implications for the quantity of food required if everyone in the world is to be sure to have enough to eat. Currently 800 million people go hungry every day. Demand for food means the quantity and quality of food that people are able to buy. This is not the same as the total need for food as many people cannot afford to buy the food they need. Total production by farmers is called the supply of agricultural products. The supply of food is produced by the combined efforts of farmers and of business people who provide inputs and help transform farm products to foodstuff that meets consumer demand.

The quantity and quality of agricultural supply depends on natural and market conditions, the costs of producing more of it, and the price farmers receive when they sell it. When product prices are high, farmers can afford to spend more money on producing it. They grow more, at higher costs. When prices are low or fall, farmers produce less of the lower priced product and more of something else that is more profitable. The weather, technological advances, and changes in the prices of inputs, outputs and alternative products affects the supply of agricultural products. However, when price changes, the supply of agricultural products does not change quickly. It is not possible to change rapidly what is growing on the farm.

The main implication of growing world population is that the supply of food has to increase at least at the same rate as the population grows, simply to maintain the current balance between total supply of and demand for food and current food supply per head of population. And, as population grows, farmers have to improve efficiency of production to increase the supply of food.

Food demand

One in every six people in the world does not have access to enough food to live a healthy life, or to have food security. The Food and Agriculture Organization expects it will take until 2030 to reduce to 400 million the number of people who do not have food security. In 2020 it is expected that 132 million children under six years old will be malnourished.

There are two parts to the problem of people going hungry: (1) too little food relative to the number of people who want it, and (2) the people who are hungry are not able to buy food. Shortage of income causes the shortage of food. About 1.2 billion people in the developing world live in total poverty. This means they have access to $1 per day or less for food and other needs. The poorest people in the world live in the rural parts of their country.

Since the Green Revolution of the 1960s, the world food problem has been the shortage of food available to people in poor nations. More accurately, the world food problem has been the world income problem. Farmers around the world grow enough food to feed adequately the current population of the world. The food supply problem occurs because the amount of food people in poor countries can afford to buy is less than the food they need. Wealthy people can afford to buy too much food, and many do so to the detriment of their health.

In wealthy countries, people have small families, and live long. These populations are stable or declining. The developing world has relatively high birth rates and falling death rates, and rapidly growing populations. As a population becomes wealthier the birth rate falls. Lower rates of growth in populations and higher standards of living happen together.

People in wealthy countries spend only a small proportion of an extra dollar of income buying agricultural products. People in wealthy countries spend more of increases in income on non-food products, and on **marketing** services associated with food, than they spend on more of the farm product. If a person's income in a wealthy country increases by $10, they will spend only around $1 to $2 more on food. A poor person in a poor country getting $10 more income will spend an extra $4 to $6 on food. For some products, such as starchy staple foods, increases in people's incomes lead to a decline in these products they wish to buy.

Increase in population and increases in income are the dominant causes of increases in demand for agricultural products. As people in wealthy countries become wealthier, total demand for agricultural products grows slowly. Further, with growing wealth, in developed countries agriculture plays a smaller role in the economy because the people spending their additional income on non-agricultural goods and services means the non-agricultural sector of the economy grows faster than the agricultural sector.

The critical question is whether the rates of increase in the supply of agricultural products around the world in the past fifty years will be able to be maintained in the future. In the 1970s the population of the world grew faster than 2 per cent per annum. This has since fallen to an average rate of growth of 0.04 per cent p.a. in the developed countries, and 1.7 per cent in the less developed countries. In developed countries agricultural output has grown at 2 per cent per year almost continuously since the 1950s.

In the latter half of the twentieth century, population all over the world grew more rapidly than ever before. At the same time the total supply of food increased even more rapidly, and calories supplied per head of population increased. Growth in supply outstripped growth in population and food per capita increased.

Less than 20 per cent of the growth in agricultural production over the past fifty years has come from using more land. Most of the increase in agricultural production came from better systems of production, better crop varieties, using chemicals such as herbicides and medicines for animals, more and better fertilisers, and better management of the technical, financial and risky elements of farm systems.

As people in poor countries get wealthier not only do they demand more and different food, their farmers produce more and different food. Agricultural output grows even more rapidly in countries that are moving rapidly from relatively low standards of living to higher levels of development. In the past decade the rates of increase of agricultural output in the less developed countries like China, India and Indonesia have exceeded 2 per cent per year. In the longer term, when rates at which populations are growing start to slow all around the world as people become wealthier, there will be less need for total agricultural supply to grow at 2 per cent per year.

A combination of circumstances worldwide has led to questioning about the capacity of world agriculture to continue to meet the demands of the growing population of the world. In recent years:

- demand for food has increased as large poor countries have grown richer
- widespread droughts have reduced annual production of farm commodities and led to a run-down of stocks
- investment in agricultural research in wealthy countries has declined
- costs of farm production by efficient producers such as Australian farmers have risen, and their profits have fallen, as competing countries such as the USA and those of the European Union have encouraged inefficient farming
- poor countries have disadvantaged their own farmers by keeping food prices low for consumers
- costs of farm inputs such as energy and fertiliser and chemicals have increased markedly.

The result has been that agricultural commodity prices have increased from their lowest ever level in 2005 to relatively high levels for a short time in 2008, albeit still around half the level they were in 1960. For the past 250 years the world supply of agricultural products has increased at such rapid rates and outstripped the growth in world demand that prices have declined continuously despite occasional but short-lived upward surges.

The steady decline in farm prices has meant that farm businesses have mostly struggled to stay profitable, even though the aggregate demand for food has grown steadily. This general pattern is likely to continue. As in the past, there will be periods of commodity shortages and high prices for farm products. This, however, does not usually mean prolonged prosperity for farmers, as costs also rise as a result of high prices.

From the perspective of the future of people involved in agriculture in Australia, small gains in the capacity of poor people of the world to buy food has big implications. Changes in the type and quality of food and fibre required in countries where income is growing rapidly mean that there will be demand for increased quantity and better quality of food and fibre. Australia is a small supplier of agricultural products on world markets. Small changes in the total demand for food and fibre by the large number of the world's poor, can amount to large changes in food and fibre production and exports for Australia.

Still, increases in demand for quantity and quality of food and fibre in countries that are becoming wealthier are met first by increases in farm production in the growing economies, and second by other competing producers from the developed and developing world. Increases in supply from around the world will meet a growing ability of people around the world to buy food and fibre.

Science and economics in managing productive systems

Systems

A key objective of this text is to engender appreciation that whatever phenomena we observe they are the result of a myriad of complex processes: causes and effects, actions and reactions. The process of looking at the whole of a phenomenon, and the parts comprising the whole, is known as a systems approach.

The 'whole farm' or 'farm management' approach used in this book is an interdisciplinary approach. Interdisciplinary methods are 'systems' methods. When walking over the farm with the farmers, it is impossible to avoid confronting the human, scientific and economic aspects of the situation. In this book, we draw the boundaries of the system more broadly than is often the case with approaches designated as 'agricultural systems'. The boundaries of the agricultural system in question range from the behaviour of atoms to improving paddock productivity to finances of the farm business to agribusiness marketing and trade to government policy on agricultural resource use.

A systems focus in agriculture can encompass the whole phenomenon from the international or national economy down to the behaviour of molecules in living organisms. The appropriate level of system to look at, and the appropriate mix of disciplinary knowledge to draw on to understand the parts and the whole, depends on the question. This is why analysts of agricultural phenomena sometimes say, somewhat mysteriously, 'the question is the answer!' This statement refers to the importance of identifying the problem (question) correctly. Identifying the correct question, the appropriate level of analysis of the system and appropriate balance of disciplinary knowledge to answer the question well, becomes clear.

Science has a long association with agriculture. Many branches of science originated in investigations related to questions arising in practical agriculture. Much of Australia's agriculture is possible only because of the contributions of science to understanding agricultural processes, and the ability of management to apply that understanding to increase productivity.

Agricultural science aims to reduce a problem or solve a problem. That is, relevant research tries to solve problems: either current or potential problems. Studying how agricultural commodities are produced and transformed into the products that consumers need and want involves drawing on knowledge from many basic disciplines and sub-disciplines: for example soil science, animal science, pasture science, crop science, engineering, agricultural economics, management, marketing, mathematics, politics, chemistry, physics, biochemistry, microbiology, computing, sociology.

The nature of intellectual enquiry into subjects means that it is difficult to know more and more about more and more. That is, an imperative of progress in disciplinary knowledge is that researchers narrow their focus and learn more and more about less and less. They specialise in knowing a lot about how a part, or a few parts, of the world work. There is no single cause or explanation for any phenomenon. In commercial agriculture, solutions to problems usually have technical, economic and human dimensions of varying proportionate importance, depending on the nature of the problem.

Inevitably, people looking to answer questions and solve problems in agriculture focus on those aspect(s) that they know most about. Thus, in agriculture, for instance, there are different perspectives that can be brought to bear when trying to answer questions and solve problems. One thing is sure: problem solving in agriculture requires drawing on a range of disciplines.

Science is the intellectual procedure involved in tracing cause and effect relationships to a set of reliable rules that hold against stringent tests in the context or environment investigated. In the systems that curious agriculturalists seek to understand and explain and manage, the cause and effect relationships are usually complex, with many interacting forces or processes.

Separating these chains/webs of multiple cause and multiple effect relationships demands systematic approaches. One approach is to simplify in an attempt to exclude factors other than those investigated. Simple but not simplistic! This is the traditional scientific method, called the reductionist approach. Another approach is to investigate large parts of production systems by making large numbers of accurate measurements of the performance of various components. The system is recreated 'on paper', and is known as 'modelling' or 'systems simulation' or 'budgeting'.

Regardless of whether reductionist science or systems simulation approaches are used, useful results depend on accurate measurement and sound reasoning in the analysis. The results most interesting to farmers are new technologies that reduce the impacts of factors limiting the productivity of the resources they manage, and produce the same output with less input or more output with the same input, or, even better, more output with less input. Achieving these outcomes 'shifts the farm or **activity** production function' or 'gets onto a new farm or activity production function'. Results from reductionist research—good science—provide ways to overcome the effects of limiting factors and maintain profitability in farming.

Science and the scientific method are dedicated to tidying up concepts as near as is humanly possible, and applying the best available previous knowledge to develop and test information about possible causes and effects. This is to see if conclusions can be believed, and to see if apparent 'findings' are genuinely new additions to knowledge. If so, the conclusion is not so much the 'right' or final explanation; it is just the best explanation at the current time. Later, with new information and more knowledge, new and better explanations evolve.

Progress comes from continuing to challenge the concepts, laws, rules, and models that derive from past scientific enquiries. New technologies emerge as applications of new knowledge directed to a specified purpose. Applying new technologies reveals information about the validity of the theory developed about them. Science is a formal, systematic, cumulative, and imperfect approach to answering the questions: 'Why did this happen? What caused it? What will be the effect?'

Progress in scientific understanding comes from scientists searching for alternative explanations or concepts that are consistent with all the existing observations and measurements. When one set of explanations is applied to a question and then the expected outcomes based on existing knowledge do not happen, new theories and explanations are needed.

Economists are concerned that the relatively limited resources of land, labour and capital in the world are used to satisfy as many of people's needs and wants, now and in the future, as is possible with the know-how of the time. As there is not enough land, labour and capital in the world to go around and to provide everything for everyone, then choices have to be made about what is to be produced, what resources are to be used, how, who gets the resulting output, and when they get it.

In most economies nowadays most choices about using resources are made by people going into markets to buy the resources they need to make goods and services, and to sell the goods and services that they produce. Where markets are unable to achieve good results, governments get involved in making these choices too.

The study of economics is different to the study of science. People's behaviour cannot be held constant and measured precisely and compared and predicted, as can be done with controlled experiments in science. Like scientists, in order to say something sensible about a situation, economists have to abstract from some of the complexities of the real world to the key elements of a situation.

Useful economic analyses in agriculture are built on sound technical foundations. To carry out sensible analyses, reach sound conclusions and give useful advice, agricultural economists have to understand the technical processes that underlie the question they are investigating. A truly unproductive approach to answering questions is to approach the issues as if everything is of utmost importance to analysing and solving the problem *except the technical aspects, or, except the economic aspects.* The moral of the story: to be good at agricultural economics, study agricultural science too; and to be good at agricultural science, study agricultural economics too.

Emphasising 'the whole' in agricultural production means starting with the people, applying farm-level science and economics, and going beyond the farm gate. The intricate connections between the farm business and farm life means the human aspects cannot be 'assumed out of the question': people are central to answers to any questions in agriculture. At other times boundaries around the sphere of interest are appropriate: the crop system, the animal system, the feed system and so on. It is necessary to do so to avoid the trap of being unable to say anything about anything unless you can say everything about everything.

Emphasising the **whole farm approach** means recognising that the goal of farm systems is to help meet farmers' goals. A recurring theme in this book is that while achieving high levels of scientific or **technical efficiency** is fundamental to farmers achieving their goals, maximising scientific or technical efficiency is not 'the goal'. This is because maximising scientific or technical efficiency in a production period is not the same as being economically efficient, making the most net benefit or profit, or achieving more of a mix of goals in that or any time period.

Management

Managers cannot manage what they do not understand and do not measure. Good management of resources requires empirical information and sound knowledge about processes. To manage agricultural resources in productive, profitable and sustainable ways, it is necessary to understand farming processes. The first step is to recognise that farm businesses consist of many different components that combine to form the whole farm system. All farm systems comprise a biological component, an economic component and a human component that are closely interrelated and that interact; and in each case the biological, economic and human components of the system are unique.

The success of a farm business in achieving the goals of the owners depends on both the technical production efficiency and the **economic efficiency**. Economic efficiency is profitability. Technical production efficiency and economic efficiency result from the way the managers of the farm family manipulate the interacting biological, human and economic parts of the business in order to achieve their objectives.

To solve problems in a farm system, professional agriculturalists first have to be able to diagnose problems correctly. To do this they need to appreciate how the farm business works as a whole. Once the problem is identified, the sensible solution(s) then often reduces to a few clear choices. Thus the key to solving problems in agriculture, and to making decisions about farm management, is to appreciate all the dimensions of the operations of a farm business.

The technical manner in which the farm family manages the resources of the farming business has a huge impact on whether battles for profitability are won or lost. In economic terms, one farm may be earning returns on capital each year of 10 per cent per annum, while another might be earning only 1 per cent per annum return on capital. Such differences in production and profitability exist between similar adjacent farms producing similar products using similar systems of production. The question: 'If so, why?', or 'If not, why not?' Why are the resources of the low-return farm business not producing the higher returns that other farms are achieving? The answer is complex, involving as it does differences in physical resources, access to capital, quality of human capital, management, attitudes to **risk**, stage of life of members of the farm family, and their goals and objectives.

In essence though, successful managers achieve their aims by making good decisions (plus some good luck), based on sound management after productivity-increasing changes have been made. The reality in Australian agriculture is there are large differences in the quality of management and of management decisions, resulting in discrepancies in efficiency (even after allowing for differences in luck), and large differences in the extent to which goals are achieved by managers, even in similar environments and production systems. In farming, there is always much scope for better management, for achieving greater goals, and for revising ideas of 'what is' and 'what could be'.

The essence of management is to put together all the components that make up the farm system in ways that achieve as many of the objectives of the manager as possible, given the reality that objectives have to be traded off against each other, that is, in business, as in life, one cannot have everything. The essence of putting all the components of an agricultural production system together and manipulating these components to achieve desired ends, is making good judgments about how the whole system works, and how it is likely to work after changes have been made.

Results from scientific analysis of how farming systems work is information to be analysed and assimilated usefully by managers when they form the judgments they use when making decisions. Judgments derive from the assimilation and analysis of new information and information from the past (lessons from experience). Good judgments are the product of information and experience and are an integral part of making good decisions. Good fortune helps as well.

Productivity

The central tenet of this book is that the productivity of resources used in agriculture has to continue to increase, for two main reasons. The first is that increases in productivity are required because a consequence of economic growth in the economy over time is that the demand for resources from the rest of the economy causes farm input costs to increase at a greater rate than farm product prices. At the same time farm product prices tend to decline as supply increases more rapidly than demand. Therefore, greater output per unit of input, and lower costs per unit of output, is essential to sustaining the contribution that agriculture can make to the well-being of people in Australia and throughout the world.

The second reason, related to the first, is that if Australia's export-earning rural industries are going to be able to help solve Australia's persistent balance of payments problems, and help meet growing world demands for food and fibre, they will have to do so by producing food and fibre as cheaply as their international competitors. By doing so, the requirements for food and fibre by the rapidly growing world population will be met, and Australians will be able to continue to import the goods and services that they need and want.

There is no sense in achieving the necessary productivity gains in a way that jeopardises the continuing productive capacity of agricultural resources. Nor is it beneficial to achieve individual farm productivity gains in ways that cause irreversible damage to the environment and impose costs to society. There have been many cases in the past where the aims of improving productivity and maintaining resource quality have been incompatible. This situation is avoidable with sound management. Modern communities will not accept irreversible degradation of the resources used in agriculture.

There is a huge challenge facing farmers and scientists alike to develop farming practices that can deliver the regular productivity gains that are needed, without degrading biological resources and processes, or raising costs on the farm too much, or causing environmental damage and costs off the farms. The key to farming profitably and sustainably is to understand the production systems and to monitor the impact of changes to practices. In the future, the 'environmental audit' imposed on the impact of a changed practice will rank alongside its ability to improve production efficiency.

REVIEW QUESTIONS

1. What evidence is there that world food supply has increased faster than world food demand over the past fifty years?
2. Discuss the question 'What factors will determine if world food supply will keep up with increased food demand around the world in the future?'
3. Is the world food problem a problem of food supply or a problem of poverty?

FURTHER READING

Food and Agriculture Organization, *The State of Agriculture and Food*, Rome, 2000.

Runge, C. F., Senauer, B., Pardey, P. G. and Rosegrant, M. W., *Ending Hunger in Our Lifetime*, International Food Policy Research Institute, Stanford, 2000.

WEBSITES

Australasian Agribusiness Review and Australasian Agribusiness Perspectives: www.agrifood.info/

Australasian Farm Business Management Network: www.csu.edu.au/faculty/science/saws/afbmnetwork/

Australian Bureau of Agricultural and Resource Economics: www.abareconomics.com/

Australian Farm Institute: www.farminstitute.org.au/

Intergovernmental Panel on Climate Change: www.ipcc.ch/

National Farmers Federation: www.nff.org.au/

World Bank: www.worldbank.org/

PART 2

Resources and Farming in Australia

Agricultural Resources of Australia

2

CONTENTS

Introduction

A resource is something that is directly useful to people or has value in some other way. What happens where, when and how, in what form, for whom, in agriculture in Australia is determined by the mixture and interplay of natural, economic and human resources and forces.

The resources farmers use in their agricultural activities are air, light and heat, land, water, capital, labour, knowledge, technology, management skill and their ability to take risks. Air is oxygen, carbon dioxide and other constituents that living organisms use. Light and heat is solar energy. Land is soil and its constituents, plants and animals living on the land, and topography. Water is rainfall, the flows of streams and rivers, water in lakes and dams and underground, and the seas. Capital is equipment, tools, farm improvements and structures, physical infrastructure and facilities for transport and communication. The production and management skills of farmers and their willingness to take risks are vital resources. Management is the process of deciding what to produce, how to produce, and how to sell the product. People in management and workers organise and combine the necessary resources to make and market the products. People in management take and manage risks.

The quality and quantity of the natural, capital and human resources in agriculture and horticulture continually change. These changes in resources happen partly from people's actions, and partly independently of things people do. Farmers change the state of soils by changing the types and mixes of fertilisers and plants and animals. They take water from surface streams, rivers and dams, and from underground stores and flows of water. They use and add to stocks of capital and technology. Farmers gain experience and learn new skills.

Important questions about resources are:

1 What constitutes a resource at any time?
2 Who owns the right to use them?
3 Are they being well maintained, being changed for the worse or being used up without renewal; and does it matter?
4 Does the way resources are used on farms affect other people or other businesses elsewhere?
5 Are the resources valuable for current or future use, or are they of value for their own intrinsic merit?

People own and control resources individually or collectively. Some resources used in agriculture are renewable and some are not renewable. Fertility of soil, water and water quality, forest timber, vegetation, animals, equipment, and knowledge are renewable resources. Coal, oil, natural gas, minerals, soil, and fertilisers such as phosphorus are not renewable.

Some resources are valued because they are useful now, such as oil, coal, natural gas, and marginal agricultural lands. They may or may not be needed further into the future. Plant and animal species are valued for intrinsic merit now and forever. They are part of total plant and animal gene pools.

The quantity of something produced from using resources is called productivity. Another term is efficiency. Technical (or physical) efficiency is the physical output produced from resources used. Economic efficiency is being able to produce under conditions of free market competition, selling products internationally, and without subsidies. In this respect most Australian farmers use their resources efficiently.

Climate in Australia

Australia is a large land mass that is dry and wet in different parts. Climate, especially rain and temperature, have large effects on whether or not agriculture is profitable. Evidence about the changing climate of Australia is found in natural phenomena going back thousands of years. There are more than a hundred years of measurements and recorded information about rainfall and temperatures that have occurred in Australia. Scientists who have looked at all of this information have concluded that the planet is warming and greenhouse gases resulting from activity of people are contributing to the warming effect. Scientists believe:

- parts of Australia have always been getting hotter or getting colder
- parts of Australia have always been getting wetter or getting drier
- in the next hundred years Australia is likely to be getting hotter
- in the next hundred years Australia is likely to be getting drier in the south, and wetter in the north
- in the next hundred years weather conditions in Australia are likely to become more volatile than in the past hundred years.

Scientists are projecting that by 2030 annual temperatures averaged across Australia will rise by 1°C above the 1990s temperature and possibly more than this, and that temperatures will probably increase by several degrees by 2070. The projected warming of Australia will vary significantly from place to place and from time to time, for example in mountains, coasts and arid inland, and in winter, spring, summer and autumn. What happens to climate in the longer term is likely to depend in part on what is done about the emission of greenhouse gases all over the world.

Projections are that rising temperatures will mean that by 2030 rainfall could decline by 2–5 per cent in winter in southern Australia. These changes are expected to be more than this by 2050 and 2070. For example, the really dry years of 2006–07 in southern Australia could be similar to 'the norm' types of seasons in 2050. It is also predicted that the intensity of rainfalls will increase and the number of dry days will increase too. A significant result of declining rainfall is declining run-off of water into rivers, streams, storages and to underground. Declining run-off means less water available for irrigation.

In many ways, for farmers who have long battled nature in Australia, drier and hotter times in the future will be more of what they have confronted intermittently throughout their farming lives. They have always confronted long and short droughts, and intermittent floods, of varying severity. Generally farmers have adapted well to change in the past, stayed in business for long times, and mostly accumulated wealth at least comparable to other sectors of the population. Global warming will be another test of the well-established ability of farmers to adapt to changes in the conditions in which they farm. Most will adapt successfully. They will change their systems and remain profitable. The capacity of farmers to adapt to change is helped greatly by good science and good thinking about solving the new problems they encounter.

Rainfall

The time of the year it rains influences what can be produced where, when and how. The rainfall of Australia is characterised by great variability, with summer maximum rainfalls in the monsoon areas of the tropical north and winter maximum rainfalls in the temperate south-west and south-east. Eighty per cent of the land receives less than 600 mm of rain each year. Fifty per cent of the land receives less than 300 mm of rain each year. The dry centre receives less than 200 mm of rain each year.

From the arid centre of Australia towards the southern, south-western, eastern and northern coasts, annual average rainfall increases steadily. First is the semi-arid region that receives slightly more rainfall per year than the arid centre. Moving towards the coasts to the east, south-west, south and north the annual rainfall increases. Beyond the semi-arid zone and inland from the coast, rainfall is 200–400 mm per year.

In the coastal regions of the temperate south and south-eastern Australia, there is usually 400–800 mm of rain per year. Midway to the southern, south-eastern and south-western coasts 200–400 mm of rainfall is received. Closer to the coast, and along the coast, 400–800 mm of rain falls annually.

Moving further northwards and north-east from the centre are the wet tropics and sub-tropics. Along the east coast north of Sydney, 800–1200 mm of rain is received per year. In the tropics of the far north, there is more than 1600 mm of rain per year. In the north-west of Australia are the dry tropics and sub-tropics.

Apart from rainfall being low, it is also highly variable between years over most of the land. Annual rainfall in the semi-arid and arid inland varies around the long-term annual average by 20 to 40 per cent. Annual rainfall in the major agricultural areas varies around the long-term average by up to 20 per cent p.a. Only parts of the south-east, south-west and the far north reliably receive rainfall that is similar in most years. The variability of annual rainfall does not indicate the impact of variability of the timing of rainfall within the year. Timing of rain matters. Rain is most useful when crops and pastures are able to make best use of it.

Unexpected dry periods and droughts are a frequent and inevitable part of agriculture in Australia. Most farmers have strategies for drought. The Bureau of Meteorology defines a drought as being when rainfall for a region has been at or below the level of the lowest 10 per cent of rainfalls known in history for a period of three or more months. In the past hundred years there have been major droughts in the eastern states of Australia in 1888, 1902, 1914/15, 1944/45, 1967/68, 1972/73, 1982/83, 1994, 2003 and 2006. The last three major droughts have each reduced the national annual gross domestic product by 1 per cent. Each time taxpayers have contributed around $1 billion to help farmers.

Australia has low average annual run-off from rainfall into streams and rivers and water storages. Average annual run-off is 420 mm. The average annual run-off around the world is 660 mm. However, as Australia has a low population the water available per person is good compared with the rest of the world. While all flows of water that can be diverted to irrigation uses in southern Australia have been diverted, little of the water in the north has been diverted for use. Most of this water is a long way from productive agricultural land, as well as long distances from populations and markets for inputs and outputs. There have been many failures in attempts to use these northern waters profitably.

Effective rainfall

Effective rain stays in the soil and plants use it. Evaporation by heat and air movement, and transpiration by plants, removes moisture from the soil. Effective rainfall is defined as rain that exceeds one-third of annual evaporation. The expected effective rainfall indicates if an area suits agricultural activities. Effective rainfall corresponds to land uses, as follows:

- no effective rainfall: very little or no agriculture
- under 5 months of effective rainfall: pastoral sheep and beef cattle, annual native grasses
- 5 to 9 months of effective rainfall: wheat and sheep, annual pastures
- more than 9 months of effective rainfall: sheep, beef and dairy cattle, permanent improved pastures.

Temperature

Temperature also determines agricultural activity. Average winter temperatures decline from the north to the south of the continent. Mid-winter temperatures during the day are 27–33°C in Darwin in the north. This declines to 21–27°C in daytime in Alice Springs in the centre. In an area bounded by Perth, Alice Springs and Brisbane, day winter temperatures are 15–21°C. Further south, from Perth to Brisbane to Hobart, day temperatures fall to 9–15°C in winter. Down south, day temperatures of 3–9°C occur in the highlands of Victoria and New South Wales and in Tasmania.

Types of climates

Australia has four types of climate: (1) Mediterranean-type climate, (2) Temperate climate, (3) Sub-tropical climate (wet and dry) and (4) Tropical climate. These climate types are defined by how much it rains, when it rains, how long plants can grow for, and temperatures.

Mediterranean climate

The Mediterranean-type climate extends from Perth across the continent to most of South Australia, western and central Victoria, to the Riverina in NSW and the lower western parts of NSW, and Tasmania. Mediterranean-type climates have wet and mild winters and hot dry summers. In Australia's Mediterranean-type climate, farmers grow grain, annual legumes and grasses. These grass and legume pastures produce feed containing protein and energy for animals during the autumn, winter and spring. Animal feed in the paddock in the summer and early autumn comes from annual legumes and grasses that have senesced and are dry not green, from the dry seed of pasture plants, from the stubble of harvested crops and from supplementary feeds supplied by farmers. The major shortages of feed in paddocks are in the summer and early autumn, and in cold, wet times of winter. The wettest parts of the country such as South Gippsland and parts of Tasmania and along the coasts can grow perennial pastures that are a mix of grass and legume species.

Temperate climate

The climate of the east coast of Australia is temperate. The temperate region is south of the sub-tropical regions that start at the latitude of 30° South. It extends from the coastal hinterlands of eastern Victoria inland and northward to the NSW Tablelands. Regions with temperate climate have an even annual distribution of rainfall and lower winter temperatures than the Mediterranean climate. In the temperate areas with high rainfall, farmers grow perennial and annual grass and legume pastures, such as ryegrass and white clover pastures. Phalaris and sub-clover pastures are grown where rainfall is lower. Shortages of feed occur in paddocks when summer or autumn is dry. Low temperatures in the winter also cause shortages of paddock feed.

Sub-tropical climate

Australia has wet and dry regions in the sub-tropics. The dry, semi-arid sub-tropics extend inland from the east coast sub-tropics right across to the west coast sub-tropics. The transition from the winter rainfall dominance of the south and the summer rainfall dominance of the north occurs in the wet sub-tropics.

Brisbane has a sub-tropical climate. The wet sub-tropical regions of eastern Australia receive reliable summer rain. Winter rainfall varies. Total annual rainfall is heavier and more reliable further north. In the wet sub-tropics, tropical grasses and legumes provide feed during the late spring, summer and autumn. Winter forage crops can be grown where winter rain is reliable. Cereals, oilseeds and cotton are grown in the drier inland of north-western NSW. Sheep and cattle graze native pastures. Sugar cane grows in the humid and wet sub-tropics.

Tropical climate

The tropical region is north of Rockhampton in Queensland and Broome in Western Australia. The tropics have reliable hot wet summers and mild dry winters. In the tropics, sugar cane, irrigated horticultural products and beef cattle are produced.

Land

The basic resource used in agriculture is land. Australia's land area is 768 million hectares. Most of it is flat and low lying. Land tenure is mostly under freehold title or some form of long-term lease from the Crown. Most land in the better farming areas is freehold tenure and inland country used for pastoralism is mainly long-term lease land. Land use in Australia is shown in table 2.1.

TABLE 2.1 Land use in Australia

		MILLION HECTARES
AGRICULTURE		465
	Arable agriculture Crops Sown pastures and grasses	**46** 16 30
	Non-arable agriculture and no-use	**419**
FORESTRY		
	Plantations	41
	Native forest	1
	Wilderness, National Parks, Conservation Reserves	37
Total		768

Forty-six million hectares suit broad-area crops, introduced pastures, and intensive and irrigated agriculture and horticulture activities. Where the rainfall is less than 250 mm in the south of the continent or less than 350 mm in the north, the land cannot be used to grow crops or introduced pastures. About 419 million hectares has such low and variable rainfall and poor soils that cropping or pasture improvement is not possible. Low intensity pastoralism is the only agricultural use possible. Large areas of land are not used for any agriculture.

Soil

The interaction of geological, botanical, hydrological, topographical and climatological phenomena, along with the processes involved in growing crops and pastures and grazing animals, determine the nature of soil at any time in any area.

Australian soils come from parent materials that have been weathered severely and suffered deep leaching of nutrients. Generally the soils do not contain sufficient of the major and minor nutrients that plants need to grow, especially phosphorus. These have to be added by the farmer. Australian soils are also low in organic matter.

Soils are less weathered in the temperate areas than in the sub-tropical and tropical high-rainfall zones. Soils in temperate areas are alkaline to acid, with some organic matter. Soils in the sub-tropical and tropical high-rainfall zones are heavily weathered and leached, and often acidic. In the semi-arid zone the soils have not been heavily weathered or leached; they are often alkaline clays with low permeability.

The total area of soils that are well suited for agriculture, and receive sufficient rainfall, is only a small proportion of the total land area of Australia. The best soils for agricultural uses are the soils that derive from basalt. Small areas of these soils occur mainly along the east coast in hilly country. The black and grey soils of north-western NSW and southern and central Queensland such as on the Darling Downs are good for agriculture and horticulture. There are some good alluvial soils along the lower reaches of coastal rivers, and in parts of the Murray–Darling Basin.

Only small areas of soils in Australia are deep and well-drained, and have the high fertility and high capacity to hold water that makes them very good agricultural and horticultural soils. Farmers generally make do with soils that are relatively shallow, have poor water-holding capacity, and low nutrient status. They are capable of achieving only moderate fertility and low plant yields, relative to the soils of their international competitors.

Even though the quality of agricultural land in Australia is generally poor, compared to overseas competitors, there is plenty of it. Fortunately soil is only one of the inputs that are combined to produce agricultural products. Soil is no more or less important than any of the other resources used to produce a farm product. This means that despite poor soils, farmers make good use of all the other inputs to production such as technology, genetics, capital, labour and management, and still make profits.

Irrigation water

The quantity and quality of water supply helps determine agricultural activities. Three-quarters of water used in Australia is for irrigated agriculture, such as cotton, fruit and vegetables, grapes, sugar, rice, dairying, pastures for intensive livestock for meat and some irrigated cereals.

The run-off of water into rivers, and the river flows, are highly seasonal and highly unreliable. The rates of flow in the major rivers, from peak to minimum flows, are extreme. Peak to minimum flows of river water can vary by 150 times, whereas variation of flows of 2–3 times is typical in most of the world. Two-thirds of the total annual run-off of surface water is in high-rainfall coastal areas above the sub-tropics, especially in the relatively unpopulated north-western WA.

Since the early days of Australia's development, irrigation for agriculture and horticulture expanded along the inland waterways, which have highly variable annual flows. The states own storages on rivers and issue licences to use irrigation water from bores or rivers. Irrigation water is distributed to farmers from facilities owned by the state, by state-owned and private irrigation companies.

There are 1.5 million hectares of irrigated land. Irrigation occurs along all the major rivers: the Murray, Murrumbidgee, Goulburn, Darling, Macquarie, Namoi, Gwydir, Barwon, Ord, Burdekin, Condamine, and on numerous minor streams. Irrigation occurs too from underground water sources and storages on farms.

Two-thirds of the irrigated farmland of Australia is in the Murray–Darling Basin. This covers parts of the Riverina in NSW and the Northern Country and Central regions of Victoria. Water comes from the Murray, Murrumbidgee and Goulburn rivers. Storage capacity in the Murray–Darling basin has increased from 5 million megalitres in 1950 to 30 million megalitres. Over 1 million hectares are irrigated in this region and on average 8 megalitres of water per hectare per year are applied. The total water used annually for irrigation in the Murray–Darling Basin is 8 million megalitres. This has increased from 3 million megalitres in 1930. Annual natural flow in the Murray–Darling river system is mostly diverted for irrigated agriculture and horticulture.

Irrigated agriculture can cause soil to become saline if more water is supplied than plants use, because excess water percolates down and raises the level of groundwater tables. Rising groundwaters bring salt upwards to plant root zones and to the surface. Downstream river water also becomes increasingly saline. Large areas of land within the region known as the Riverine plains in NSW and Victoria, covering the Deniliquin, Wakool, Murrumbidgee, Shepparton and Kerang regions, have shallow groundwater tables and are vulnerable to salinity.

In response to problems of water being used inefficiently in the economy, with increasing salinity and with insufficient water flowing to the natural environment as a result of water catchment and use for irrigated agriculture, governments have established markets for water for irrigation. The aim is

FIGURE 2.1 Australia's major irrigation schemes

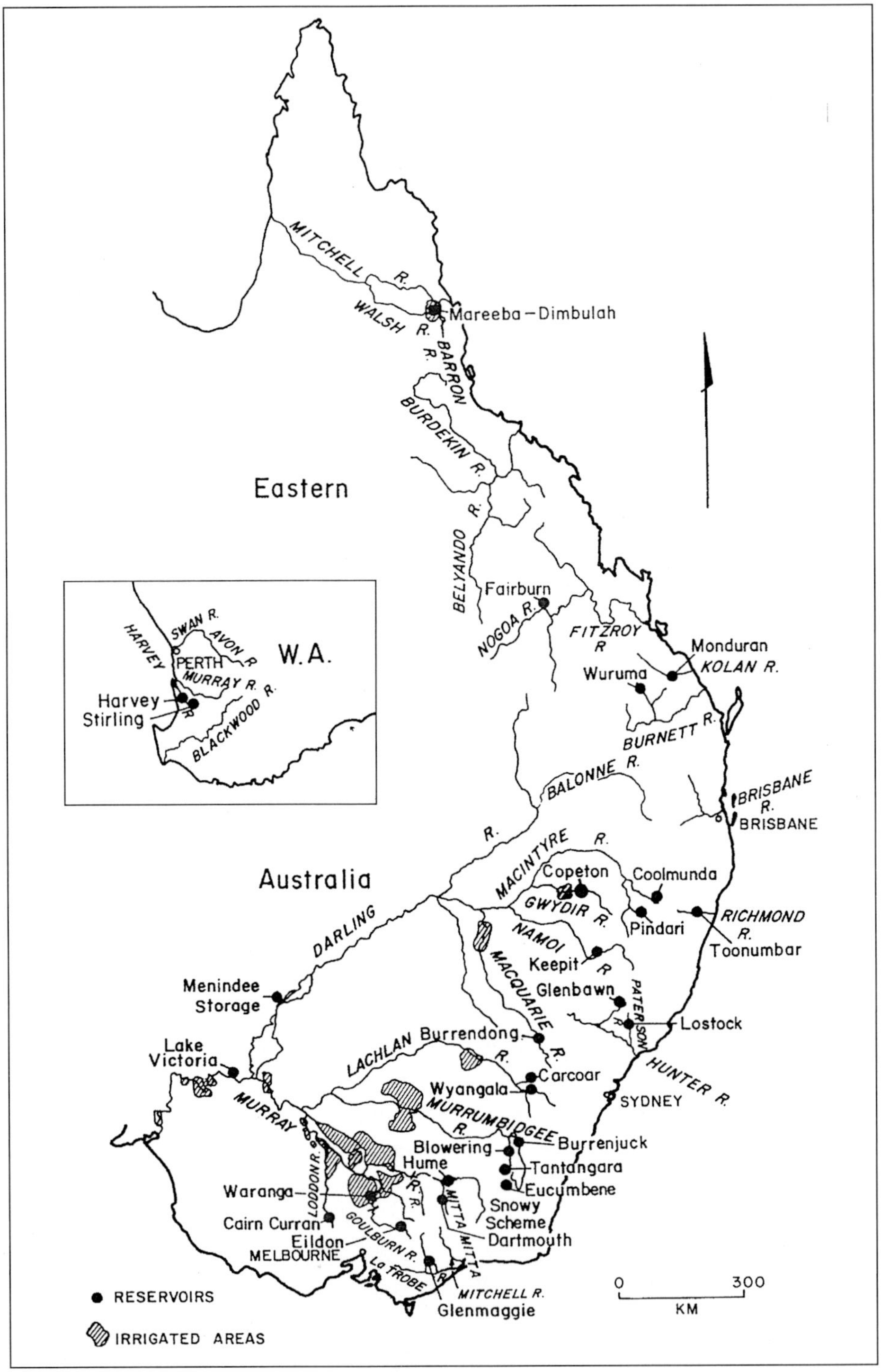

Source: Davidson, B. R., *European Farming in Australia*, Elsevier, 1981

to ensure that irrigation schemes operate more effectively and farmers, by paying the market price for water, are forced to use water as efficiently as other potential buyers and users of the water in the economy. Buying and selling water in markets enables irrigation water to be moved from less valuable uses to more valuable uses, to the net benefit of the nation.

Agricultural production around Australia

What happens when, where, why and how in agriculture and horticulture in Australia is the question of interest. Farmers farm in every state: 40,000 in NSW (30 per cent), 32,000 in Victoria (25 per cent), 27,000 in Queensland (20 per cent), 14,000 in SA (11 per cent), 12,000 in WA (9 per cent), 3800 in Tasmania (3 per cent), 380 in the Northern Territory (0.3 per cent). The main farm activities are cattle for meat; wheat, barley, grain legumes and oil-seed production, usually in conjunction with animals; woolgrowing; sheep and lambs for meat; dairying; sugar; cotton; vegetables; fruit; poultry for eggs and for meat; and pigs.

FIGURE 2.2 Australia's farming zones

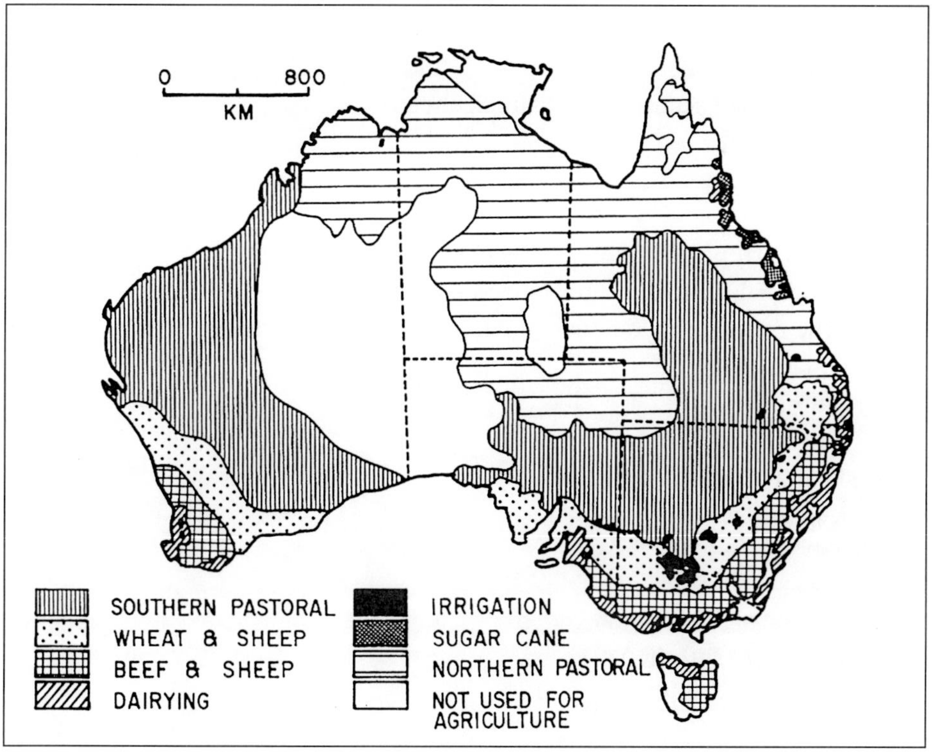

Source: Davidson, B. R., *European Farming in Australia*, Elsevier, 1981

Cereals are grown in the southern mainland states. Mixed sheep–cereal farms are in NSW, Victoria, SA and WA. Beef is produced in all states, mainly in NSW, Victoria and Queensland. Sheep meat is produced in Victoria, NSW and WA. The greatest number of dairy farms are in Victoria and Tasmania, with others in south-eastern SA, the southern Riverina of NSW and Queensland and WA. Pig and poultry meat and eggs are produced mainly in NSW and Queensland. Queensland has the most vegetable producers, though significant vegetable production takes place in every state. The sugar industry is in Queensland. Cotton is grown in NSW and Queensland. Apple, peach and pear producers are mostly in NSW and Victoria, as are nearly all citrus growers. Grape-growers for wine are everywhere. Northern Australia has beef and irrigated horticulture businesses.

Agricultural zones

Most of Australia's agricultural activities occur within three major zones: the high-rainfall zone, the wheat–sheep zone and the pastoral zone. There are 60,000 farm businesses in the high-rainfall zone, 55,000 farm businesses in the wheat–sheep zone and 5000 farm businesses in the pastoral zone

There are a number of sub-sectors within the high-rainfall, wheat–sheep and pastoral zones. Davidson (1981) defined Australia's agricultural zones as:

1. High-rainfall zone
   - Sheep–beef sector (high-rainfall sheep)
   - Dairy sector
2. Wheat–sheep zone
3. Pastoral zone
   - Northern pastoral sector (Beef)
   - Southern pastoral sector (sheep–beef).

Within the zones defined above, the following activities are located:

1. Sugar cane
2. Irrigation
   - Crops
   - Fruit (including viticulture)
   - Vegetables
3. Pigs
4. Poultry.

The high-rainfall zone

The high-rainfall zone receives on average over 500 mm of rainfall each year. This zone covers the coastal lands and adjacent tablelands in Victoria, NSW and Queensland, small areas of south-eastern SA and south-western WA, and the whole of Tasmania.

The high-rainfall zone has steep sloping land and tablelands, with good supplies of surface water. In the south, rainfall is winter-dominant. In northern NSW and southern Queensland, rainfall is spread more evenly through the year, with some summer dominance that grows sub-tropical and tropical pastures. The major constraint on growth of pasture plants is low temperatures in the winter-dominant rainfall areas of the south, and low moisture in the summer-dominant rainfall areas. Quality of feed is high in the winter in the south, and low in the winter in the summer-dominant rainfall areas. The high-rainfall regions are used mainly for grazing improved pastures. Stocking rates per hectare are high. Cereal crops are grown where soils drain well. Most dairying takes place in the high-rainfall zone.

Sheep–beef and cropping sector

Grazing of sheep and beef, and cereal cropping are the main activities in this region. The growing season ranges from six to nine months. Prime lamb and beef production is important throughout the region. Farm sizes range from the many and increasing relatively small part-time farms to large-scale operations of 5000 to 10,000 hectares. Merino woolgrowing was once a major activity but a long run of low wool prices and high prime lamb prices led to an increase in prime lamb activities and a reduction in merino wool activities.

Specialised beef cattle properties in the high-rainfall region are largely confined to districts where the terrain is too steep for dairying, and the rainfall and pasture growth are unsuitable for sheep. Stocking rates differ markedly between properties. Around 12–18 dry sheep equivalents grazing on a hectare is common. Cropping is increasing. Oats, barley and canola are grown for grazing and fodder or grain that is sold, or stored for supplementary feeding. Cereals are grown for grazing and grain.

The main native grasses in the high-winter-rainfall areas are perennial grasses such as Kangaroo grass, Red grass and Spear grass, and annual grasses such as Wallaby grass and Winter grass. Native grasses in the high-summer-rainfall areas include the perennial *Aristida* and *Chloris* species. These native grasses have useful nutritive value if managed well.

Introduced species of grasses and legumes increase total herbage production. The quality and digestibility of the feed throughout the year is raised and spread more evenly. Introduced pastures can be established over most of the sector either by cultivation techniques or by aerial seeding.

Dairying sector

Australia's dairy production is predominantly either within the high-rainfall zone or on irrigation areas such as the Murray and Goulburn and Macalister irrigation areas in Victoria. Most dairy farmers milk 250–300 cows or more. Increasingly herds of 400, 500, 600 or 1000 cows are being milked. Labour is supplied by the family plus employees. Dairying is based on grazing either rain-fed or irrigated pasture, plus conserved fodder transferred through the year plus concentrates that make up varying but always significant proportions of total feed supply.

The basis of most rain-fed dairying pastures are mainly perennial grasses and clovers. Irrigated dairying uses perennial and annual pasture species. High-pasture yields, mild temperatures and long growing seasons in Victoria and Tasmania make dairying highly productive. In the sub-tropical coastal areas of northern NSW and south-eastern Queensland, winters are relatively dry and the summer rainfall suits Paspalum, Rhodes grass and Kikuyu as the pasture species. Much more supplementary feeding is needed in winter and spring than in the more southerly dairying areas.

In the past four decades the dairy industry has been through a massive process of adjustment. The number of dairy farmers has fallen from 45,000 in 1970 to 8000 in 2008. The cows milked per person and production per cow has increased greatly.

Wheat–sheep zone

The wheat–sheep zone is the second most important zone of agricultural production. About 33 per cent of Australia's agricultural output is produced in the wheat–sheep region, including most of the wheat grown. The wheat–sheep zone has a growing season from five to nine months. The climate and the topography suits regular cropping accompanied by sheep or beef grazing on improved and native pastures. Stocking rates per hectare are lower than in the high-rainfall zone, and properties are generally larger than in the high-rainfall zone.

Nearly 75 per cent of properties produce sheep, wheat and barley, accompanied by grain legumes, pulses, and oil-seeds in the crop rotation. The remainder are sheep or sheep–beef properties. The wheat–sheep region carries half the national sheep flock. Sheep **enterprises** are Merino woolgrowing (medium fibre diameter), prime lamb production, and stud-breeding. Beef cattle are also produced in small numbers. In the northern parts of this zone, cereal crops are replaced by sub-tropical crops such as safflower and summer-growing crops such as sorghum or sunflowers.

The areas producing wheat receive from 250 mm to 600 mm of rainfall each year. Most of this rain falls in the winter and spring, though the northern NSW and southern Queensland wheat-growing areas receive the majority of total rainfall in the spring and summer. Soils used for wheat growing are generally only moderately fertile, ranging from sandy soils to heavy clays. Short annual pasture phases are grown when a ley farming system is used.

The pasture phases of crop rotations in WA, SA and the Victorian Mallee commonly are for only one to two years, based on annual pasture species regenerating from hard seed remaining in the soil. In NSW and the better areas of WA, SA and Victoria, longer pasture phases of three to four years of perennial and annual species of pasture legumes are used. The native grasses in this zone are low producing, supplemented by annual pastures of ryegrass, subterranean clover and medics. Improved pastures use annual medics in the drier areas and subterranean clover in the higher rainfall parts.

Farmers move their activities into and out of cereals and sheep to some extent as prices and net returns from cereals and sheep and lambs and wool move up and down. The pasture phase reduces when grain prices are high relative to the prices of livestock products, and the crop area reduces when sheep and wool prices are high. Increased emphasis on cropping in this zone has led to larger areas of crop, larger agricultural machinery, reduced tillage operations, more chemical weed control and intensified rotations using grain legumes, pulses and oil-seeds.

Pastoral zone

The pastoral zone covers the arid and semi-arid areas of Australia. Rainfall is low. The main land use is grazing native pastures at low stocking rates per hectare on large land areas. Properties in the pastoral zone produce around 10 per cent of the gross value of cattle production per year in Australia, and about 5 per cent of the gross value of wool produced. Costs are high because of the high costs of transport of inputs and output and the high costs caused by the riskiness of production. Non-agricultural activities such as mining and tourism are more significant to the gross domestic product than pastoralism in the arid and semi-arid areas.

Northern pastoral zone: beef sector

The northern pastoral beef region takes in the northern parts of WA, the Northern Territory, and northern and south-western Queensland (the Channel Country). The sector comprises one-quarter of the total land used for agricultural production in Australia. It produces a small proportion of gross rural output. Beef cattle production is the major agricultural activity. It is an area of great diversity in rainfall, carrying capacity and production techniques.

The far north of Australia has a wet season that is highly variable and followed by high temperatures. Animal growth occurs mainly in the wet season. Soils are poor. Grazing is based on tropical grasses of low nutritive value for most of the year, except in central Queensland and on the Barkly Tableland on the Queensland–Northern Territory border where Mitchell grass provides good grazing.

Cattle are produced with a minimum of production operations such as mustering, branding, weaning and drafting. Traditionally cattle were turned off as low-quality lean beef at four to six years of age. Better pastures, better genetics, efficient mustering using helicopters, more transporting of cattle, growth of the live cattle market and increased feed lotting for domestic and export markets has led to cattle from the northern beef zone being finished for sale at younger ages.

Carrying capacity ranges from one beast on 30 to 55 hectares. Average property size in the Kimberley area of WA is 250,000 hectares, carrying 9000 head of cattle. In the Northern Territory, average station size is 450,000 hectares, carrying 8000 head of cattle. In far north Queensland, properties of 300,000 hectares can carry 5000 breeding cattle.

The freight cost of inputs and the cost of transporting sale animals to meatworks or to buyers elsewhere are high. Labour supplies are limited. The profitability of cattle production fluctuates markedly. There are occasional years of high profits and runs of poor years.

The majority of the output is sold on volatile export markets. Most cattle-producing businesses in this area lease their land on long terms. Leasehold, as a form of tenure, has important implications for management, particularly in relation to the incentives for the maintenance and improvement of properties within the zone.

Southern pastoral zone: sheep–beef sector

The pastoral sheep–beef sector occupies approximately 25 per cent of the agricultural land in Australia and produces about 10 per cent of Australia's rural production. Most of this sector is south of the Tropic of Capricorn with the exception of a small segment in central Queensland and WA.

Almost all the land used for wool production in the arid and semi-arid areas is held under conditions of leasehold. The lessee has to provide watering for the stock and boundary fences and meet various other conditions to do with limits to stocking rate and cropping.

Merino woolgrowing is the main sheep activity. Properties are large, ranging from 5000 to 100,000 hectares or more; paddocks are 4000 hectares or more, and flocks of 5000 to 20,000 head are common. Labour and other inputs used are low relative to other agricultural enterprises in Australia and overseas. Capital equipment is limited to fences, water facilities, yards and vehicles. Despite low lambing percentages, the emphasis is almost exclusively on breeding flocks because of the cost of obtaining replacements for non-breeding activities. Sheep husbandry practices are simple: mainly mustering, marking and shearing. Internal and external parasites, apart from the sheep blowfly, are not major problems because of the hot, dry climate.

The major plant species used in sheep production are the grasses that appear after rains: the bluebush and saltbush shrubs in the southern areas; and Mitchell grass in the arid tussock grasslands in the north-east of this zone. These native pasture species are well adapted to the low, variable rainfall, and low soil fertility. They produce low dry matter and cannot withstand heavy grazing. With low and highly variable rainfall, feed supply can vary from three-fold to five-fold from year to year.

Most of the plant growth occurs over only two to three months of the year. The quality and digestibility of plants deteriorate rapidly after flowering. Stocking rates are set at levels that are able to be maintained throughout the whole year, year-in year-out. In years of good rains and plant growth, the feed supply consists of high-quality and highly digestible grasses annual grasses. The rest of the feed supply comprises perennial grasses of lesser quality, while perennial shrubs such as saltbush and bluebush provide reasonable levels of protein. In some areas the feed supply depends on regular periodic fires that help control woody shrubs and stimulate forage growth.

Drought is a perennial hazard in the arid zone, posing threats to businesses and the natural resources. The rangelands are a fragile ecosystem, vulnerable to the effects of grazing and over-grazing. Environmental problems emerging in the rangelands include the proliferation of the shrubs known as woody weeds and the dangers of soil erosion.

Intensive activities located within various zones

Sugar cane

Sugar cane is grown mainly in the north and central coasts of Queensland where around 350,000 hectares of sugar are grown. New areas of sugar are being grown in the Ord River region of Western Australia. Sugar cane is harvested and delivered to sugar crushing mills that are close to the farms. The Australian sugar cane growing industry is changing, with fewer producers and total output stable. Around 5000 sugar cane growers grow on average around 8500 tonnes of sugar cane per farm. Three per cent of growers produce around 22 per cent of national sugar cane production.

Typically, over 20 million tonnes of cane is produced in Queensland each year. This produces over 3 million tonnes of raw sugar. Eighty per cent of this raw sugar is exported, which is about 10 per cent of world trade in sugar. Australia exports mainly to Japan (20 per cent), Malaysia (13 per cent), Canada (15 per cent), Korea (13 per cent), and China (7 per cent.) Sugar production and prices vary greatly from year to year according to seasonal and market conditions. Australian sugar cane producers face increasing competition from large exporting countries such as Brazil.

Irrigation

Irrigated farmland makes up 0.5 per cent of all agricultural land, and uses 70 per cent of all the water taken from water sources for rural, industry and domestic uses. In years when there is the expected average supply of irrigation water, one-quarter of the gross value of agricultural production comes from irrigated farms. Half of this is irrigated horticulture, one-quarter is irrigated pastures and one-quarter is irrigated broad-acre crops.

Irrigation settlements also played a big part in the evolution of modern rural Australia. In the 1880s private irrigation trusts were formed, mostly to irrigate pastures. Though they had the virtues of local participation and self-help, many ran into financial difficulties. In the twentieth century, governments set up irrigation schemes for agriculture and horticulture with the aim to populate the countryside and stimulate development of the states. Large irrigation settlements were set up such as the Murrumbidgee and Murray Irrigation Schemes in the 1930s. The Snowy Mountains Scheme was set up in the 1970s. Up until the 1990s, irrigation developments were encouraged because the supply of irrigation water was thought to be ample.

The formula applied originally to Australian irrigation schemes was that the farmers cover **operating costs** and the taxpayers bear the capital costs. External costs such as salinity of land and saline river flows downstream were not significant concerns at this time. With the infrastructure for irrigation agriculture provided to farmers at no cost to them, and the users of the water charged less than it cost, inefficient use of water and external costs were inevitable. As well, until the 1980s most of the publicly provided irrigation water was used to produce commodities that required even further assistance through subsidy, or artificially high domestic prices.

Irrigation brought dramatic change to the countryside. However, the expected productivity and profitability of irrigation farming was generally over-estimated, or the calculations too simplistic. The costs of the inevitable salinity and drainage problems that surfaced eventually were not anticipated in the initial enthusiasm for irrigation farming. Also unanticipated was the high productivity and profitability of improvements in dryland agriculture that were made possible later by the superphosphate and subterranean and medic clover and trace element revolutions in the pastoral and cropping areas.

Nowadays Australian water use is characterised by high demand for water, and keen competition for it among a wide range of agricultural, industrial, urban and environmental uses. All large-scale public irrigation projects in southern Australia that are economically feasible are developed. Water storage and delivery infrastructure is wearing out and needing new investment. The price of water has increased markedly in the past decade. So too has the amount of water traded between different users and uses increased greatly. Water can be traded between alternative irrigation farms and other uses, such as environmental flows. Demand by urban populations for water to be retained in rivers to nourish the riverine environment is strong and growing.

Serious problems of water supply, not just for irrigation agriculture and horticulture but also for urban and industrial uses, have arisen because of over-allocation of water supplies for irrigation during 1980–2000. Concerns about external costs and water for environmental uses increased greatly during the 1990s. A decade or more of dry years from the mid 1990s highlighted water shortages all over southern Australia, and hastened the introduction of markets for trading water from less valuable uses to more valuable uses.

Increasingly emphasis in public policy is on how best to utilise the existing system, including buying and selling rights to water to enable reallocation of water from low- to high-value uses through transactions between farmers and between other users—urban, industrial, and environmental.

A looming problem for some irrigation farmers is that increasing demand for water in higher-value uses will mean that water for irrigation will become too valuable in many years to continue to use irrigation water in low-value activities. The real price of irrigation water increased at a steady rate over the decade past. Irrigation farmers producing relatively low-value products such as irrigated pasture for grazing animals other than dairying, or crops that have relatively high water requirements and relatively low values such as rice, will find in many years they will be better off selling their right to use the water to others.

Major activities grown using irrigation are rice, irrigated pastures, fruit, vegetables and ornamental horticulture. Rice was produced mainly in the Murrumbidgee Irrigation Area in the Riverina of NSW when water was plentiful and cheap. Irrigated pastures are grown for dairying on rivers and streams, and are also used for fat lamb and beef production. In recent years rice production has declined by 85 per cent because there has been little water for irrigation.

Australia's 17,000 horticulturalists are found near major population centres and in all the irrigation areas. Horticultural regions range from the wet and dry northern tropics, to the sub-tropical slopes and plains, to temperate highlands, slopes and plains, and along the coasts. Irrigated vegetable activities are found interspersed throughout these zones near water sources, often near large centres. Irrigated fruit activities are found on the major watercourses in the various irrigation districts, such as the Murrumbidgee Irrigation Area in NSW (Riverina), the Sunraysia irrigation area along the Murray in the Victorian Mallee and the Riverina of NSW, the Goulburn Valley of the Northern Country of Victoria, and the Burdekin irrigation area in the Central Coast of Queensland. Fruit, vegetable and nursery production are the largest horticultural activities.

The main types of fruit produced are grapes, citrus, apples, bananas, pears, peaches, pineapples and apricots. Total production of fruit and vegetables has increased by 25 per cent over the past decade. The nursery industry has grown even more rapidly and a small export of cut flowers occurs. Most irrigated fruit and vegetable production is for domestic use. Consumption of fruit and vegetables per person has increased slightly over time. The main fruits consumed are citrus fruits and juices. The main vegetable is potatoes.

Consumption of fruit and vegetables has increased in Australia over the past fifty years: fruit consumption has increased from 81 to 135 kg per head and vegetable consumption has increased from 130 to 162 kg per head. Australia produces over 150 horticulture crops across a wide range of climatic conditions. Horticulture production in Australia is a very small proportion of world horticultural output, around 0.5 per cent.

Queensland, Victoria and NSW are the main states producing horticultural products (excluding wine), and this production is mostly irrigated, using 17 per cent of Australia's irrigation water. These states produce over half of the fruit and vegetables of Australia. Most of the fruit and vegetables produced in Tasmania are rain-fed and go to the processing industry. The main vegetable crop is potatoes, followed by tomatoes, carrots and onions. These are grown in every state. Vegetable production is around 6 per cent of the gross value of agricultural production in Australia. Vegetable production is a labour-intensive activity.

There are three markets for vegetables: fresh vegetable market; processed vegetable market; and export market. Increasingly supermarket chains contract directly with large horticultural producers to supply fruit and vegetables.

Trade in horticultural products worldwide is growing, with supply increasingly coming from developing countries. Australian horticulture is 'out of season' with production from countries in the northern hemisphere. Horticultural produce from other major producers in the southern hemisphere comes from Chile, South Africa and New Zealand.

Australia's exports of horticultural products have declined in recent years and imports have increased. This reflects changes in the exchange rate that have been adverse to exports and favourable to imports, as well as increased competitiveness of low-cost developing countries. Increases in barriers to trade also restrict access to overseas markets.

Viticulture occupies 170,000 hectares and produces 1.5 million tonnes of grapes each year on average. Of this, 1.3 million tonnes goes to wine, in 60 wine regions. The largest wine area is in SA with 75,000 hectares. Wine is the nation's fourth largest agricultural export.

Grape-growing, wine production and wine exports have increased markedly in the past 20 years. This has resulted from the area planted to grapevines trebling since 1990 and yield per hectare increasing. Growth has been in the main red wine varieties Cabernet Sauvignon, Merlot and Shiraz and the main white wine varieties Chardonnay, Sauvignon Blanc and Colombard.

The number of horticultural businesses has declined over the past decade, and the scale of enterprises has increased. Supermarket chains contract directly with large producers for much of their fresh horticultural products. Horticultural activities have to deal with the perennial problem of distance and transport costs of perishable product where quality is extremely important, and with highly competitive export markets. Generally, Australian horticultural exports face increasing competition from low-cost producers on world markets. Western Australia is the main state for vegetable exports, and the main destinations are Europe and South-East Asia. The rapidly growing ornamental horticulture industry locates close to urban centres. The nursery industry in Australia has significant scope for growth, with spending on cut flowers and indoor plants rising with increases in disposable incomes.

Pig industry

The pig industry has changed markedly since the 1960s. Pig production was small and the industry was based on by-products of the dairy industry. In the late 1960s, large supplies of wheat led to expansion of the pig industry into grain-growing areas. Since then, the number of pig producers has declined, from around 40,000 to just over 2000 between 1970 and 2008. Specialisation is occurring. Three per cent of producers run over 1000 sows and have half the sows in Australia. The majority of producers have 100 sows or less, and have other farming activities as well. Annual pig meat production increased by over 130 per cent in this time. The growth in output has come from more pigs per litter, less deaths and greater weight gain slaughter rates.

Changes in consumer tastes have forced producers to produce leaner pigs too. Pig production is done in an environment where virtually everything is controlled except the cost of the feed and the prices of the pig meat. Most of the pig production is in areas where grain is produced. Australia's exports of pig meat have increased and so have imports. In the past decade, exports of pig meat have increased from \$50 million to \$200 million. Imports of pig meat have increased from \$40 million to \$250 million.

Egg and poultry meat industries

Egg production and poultry meat production are intensive industries that are highly technically efficient. Chicken meat producers are part of an integrated production and marketing chain. They produce chicken meat on contract for a small number of large processors and retailers. Chicken meat production and consumption has increased greatly in the past few decades. Poultry meat is similar to beef and veal in terms of consumption per head per year.

Poultry meat has become markedly cheaper in **real terms** over the past two decades because of great increases in technical efficiency. Traditionally broiler and breeder farms, and egg farms, have been located in rural areas close to major cities, for example the Mornington Peninsula near Melbourne, and the Hunter Valley near Sydney. As urban areas have grown, these industries have been forced to move further away. They locate near the grain-producing areas such as around Tamworth in NSW, St Arnaud in Victoria, and the Darling Downs in southern Queensland.

Feedlotting beef cattle and lambs for meat

Feeding young cattle and lambs in confined intensive areas on high-energy, high-protein rations of grains and forage for short times, to enable them to grow rapidly to reach the weight and quality required by particular markets, has become increasingly common in animal production in Australia.

Feedlot operations are found as part of mixed farming businesses growing grain and animals, and as separate businesses located close to grain-producing areas and close to major markets and shipping points for export.

Grain-finished beef and lamb have become a significant and stable proportion of total beef and lamb supply in Australia over the past decade. There is always an opportunistic element to feedlotting: when grain prices and meat prices are appropriate, more animals are fed grain to grow rapidly and finish with the required amount of muscle and fat cover. At times of relatively low meat prices and high grain prices, feedlots can be operating at less than half capacity.

As an intensive animal activity, environmental and animal welfare issues are important and feedlots have to meet stringent waste management and animal welfare criteria.

REVIEW QUESTIONS

1 What is a resource?
2 Technical and economic factors determine the types of farm activity that are carried out and where they occur in Australia. Identify the main determinants in the location of activities.
3 What proportion of the Australian land mass can be used for arable agriculture?
4 List the zones of agricultural production and describe where they are.
5 What are the main climate types in Australia? List the main characteristics of each climate type and the type of agriculture that can be carried out in these climates.
6 What are the main agricultural activities found in the major zones of production distributed around Australia?
7 What characteristics of Australia's climate, land and water supply dictate the type of agricultural activity and where they take place?

FURTHER READING

Cocks, K. D., *Use with Care*, University of NSW Press, Sydney, 1993.

Davidson, B. R., *European Farming in Australia*, Elsevier, Amsterdam, 1981.

Reid, R. L. (ed.), *The Manual of Australian Agriculture*, Butterworths, Sydney, 1990.

WEBSITES

Australasian Agribusiness Review and Australasian Agribusiness Perspectives: www.agrifood.info/

Australasian Farm Business Management Network: www.csu.edu.au/faculty/science/saws/afbmnetwork/

Australian Agricultural and Resource Economics Society: www.aares.info/

Australian Bureau of Agricultural and Resource Economics: www.abareconomics.com/

Australian Farm Institute: www.farminstitute.org.au/

CSIRO: www.csiro.au/

National Farmers Federation: www.nff.org.au/

NSW Department of Primary Industry: www.dpi.nsw.gov.au/

Vic Department of Primary Industry: www.dpi.vic.gov.au/dpi/index.htm

3 Farming

CONTENTS

Introduction

Farming is the business of organising and combining and re-organising people and natural resources. Farmers combine land, sunlight, rainfall and irrigation water, plants and animals with feed, fuel, fertiliser, chemicals, electricity, labour, management skill, specialist knowledge, capital equipment, financial capital and time to produce agricultural products. Doing this, more often than not, over many years, farmers achieve their goals. They do this despite farming being highly uncertain and risky.

Farming in Australia plays a major role in earning export income. Farming has done this almost from the start of European settlement. Relative to most countries, Australia has large areas of clear, flat and watered grazing lands. Consequently, from European settlement, agriculture, mainly grazing activities, developed rapidly. Early agriculture was based on using low amounts of labour and relatively low amounts of capital, added to relatively large areas of land.

Forty-six million hectares of Australia is suitable for arable agriculture. This is 6 per cent of total land. This is a substantial area of arable land for agriculture and horticulture for a small population, compared with other countries. A further 400 million hectares (60 per cent of total land area) of Australia's total 768 million hectares is used for agriculture that does not require cultivation. The remaining one-third of Australia has no agricultural use.

Having large areas of agricultural land per head of population has been, and has remained, the principal advantage farmers in Australia have over farmers in other countries. Nevertheless, compared to that of international competitors, farming in Australia is a highly risky endeavour. The key feature of the natural environment that affects farming is rain when plants are growing. This varies greatly from one year to the next. The supply of water for irrigation from rain that runs off the land and into catchments and underground is limited and highly variable. The soils generally lack natural supplies of nutrients for plants, have little organic matter and poor structures.

Australia's 120,000 small, medium-sized and large farm businesses produce for small domestic markets and large export markets. Generally farms are located a long way from their markets for inputs and outputs. The competition faced by Australia's farmers from other farmers around the world, on domestic and export markets, is fierce and increases continually.

Farmers also compete with businesses in the non-farm sector of the economy for the nation's limited supplies of land, labour, water, research funds and skills, and capital. The non-farm sector of the economy grows more rapidly than the farm sector and competition for resources increases. This means the prices farmers must pay to obtain these resources rise. At the same time the prices farmers receive for their products have fallen steadily over time, and have done so since settlement began. The two effects—increasing costs paid and falling prices received—mean that farmers have to continually improve to stay in business. They have to get more from the same resources. Most do this well and increase their productivity better than most other business-people in the economy.

In some ways the essential nature of farming has changed little over thousands of years, but at the same time farming as a business changes continuously. Farming in Australia is increasingly falling into several distinct categories. One category is 'very large-scale commercial farms'. Most of the very large-scale commercial farms are owned by a family partnership or company or trust, and occasionally by a corporate entity involving non-farming shareholders who supply the capital. Large-scale commercial farms can involve tens of millions of dollars of **capital investment** and earn millions of dollars of income. These businesses can earn profits (rates of return on capital invested) comparable to other large businesses in the economy. In another category are commercial family businesses that operate as sole proprietors, partnerships, companies or trusts and are significantly above the average size. These too involve million dollars of capital invested and earn competitive returns to capital.

The 'very large-scale commercial family farms' and 'significantly above-average commercial family farms' make up 20 per cent of the total number of farms. These farms contribute 80 per cent of annual total value of agricultural output. Another indicator of the distribution of relative shares of total farm production is that the largest 10 per cent of farm businesses account for 50 per cent of total farm output.

The next couple of categories of farms are typically small to medium-sized commercial businesses owned by a family. These farms make up 80 per cent of all farms and contribute 20 per cent of the annual total value of agricultural output. The medium-sized operations have total capital invested of $2–4 million. Another measure of the relative contribution of farms to total output is that 50 per cent of farms contribute 10 per cent of total output. Operators of the small farms struggle to earn enough profit to expand and stay in business. They need income from other sources too. A significant change in modern times has been that on average people involved in small to medium-sized farm businesses earn substantial proportions (30–40 per cent) of their total income from off-farm sources. This income comes from investments off the farm or work by family members off the farm.

Another category is farms owned by people who derive all or most of their income from doing something other than farming. These people own farmland for reasons to do with enjoying rural lifestyles. These farms are generally small but can be medium sized. Recreational farms predominate in areas of the country that are within reasonable proximity to major population centres and have attractive natural environmental features. This category of owners of farmland is growing rapidly in all states of Australia.

Farm family goals

Farmers farm because it is how they like to live and to make a living. A main reason a commercial farmer, a commercial farming family, or a commercial farm investor combines the resources they control into a farm business is to make some profit, so that they can keep on farming. Understanding how a farm business works requires understanding how all the parts that make up the farm business fit together and affect each other. This requires keen appreciation of the motivations of the owners and managers of the resources used in farm businesses.

Whenever a group of farm families are asked why they do what they do, and what it is they hope to achieve or get out of farming, the list of answers invariably includes the following:

- farming is a good way to live
- it is an exciting industry to be in
- it is a genuine intellectual and practical challenge

- it is a way of building up greater **net worth** over time
- it is a way of being part of a community with similar interests
- it is doing what we are best at and like doing best, and a good way to rear a family
- we would like to be good at something and be recognised by our peers as being good at something.

Capital

Australia's agricultural sector comprises mostly family-operated farms, with very few farms operated as public or large family companies. Around 65 per cent of all holdings operate as sole operators, around 30 per cent are run as family partnerships and around 5 per cent operate as family companies or trusts. A small proportion of companies are large farm businesses owned by non-farmers. Some of these are part of a vertically integrated production and processing operation such as are found in the horticultural industry, broiler industry, and in egg and pig production.

The majority of farm businesses are medium sized and involve a capital investment of $3 million to $6 million. The total capital of a farm business is the value of all the things used in the business: the land, animals, machinery, equipment, vehicles, tools, buildings, stocks of fertiliser, chemicals, feed on hand, and cash and other accumulated savings that are part of the business. A few farms are part of a large-scale company farming operation. For about one-third of farms nearly all the capital invested in the business belongs to the owners of the business. For about two-thirds of farms, the total capital is made up of the farm owner's capital plus some capital borrowed from others, such as bankers or relatives.

The net worth (or **equity**) of the owners of the business is the value of total capital minus the debts. For example, the assets of a farm business (the land, animals, machinery, cash) might be worth $4 million. There might be $500,000 of borrowed capital, called debt. The farm owner's equity (net worth) is $3.5 million. This is 87.5 per cent of total capital. For most farms, the net worth of the owner amounts to 75–85 per cent of the total value of the business. Commonly, farms with more debt and less equity than this find that the profit from the business does not always leave enough cash to pay the **interest** on the capital that has been borrowed. This is why most farm businesses have a high proportion of equity and a relatively low proportion of debt. However, to be successful it is necessary at various times in the life of the business to make use of a reasonable proportion of other people's capital as well as the owner's capital to enable the business to grow.

Variability of farm income

Drought is one of a number of risky and uncertain eventualities that affect cash flows and asset values in any year. Other major risky happenings that adversely disrupt planned cash flows and net worth include market collapses, government decisions, flood, fire, disease, plague and pestilence, input shortages or failures, input price rises, mechanical failures, marketing disruptions and product losses, and farm-family accidents and illnesses. Consequently, one of the most notable features about Australian agriculture is that **net farm income** varies greatly from year to year. Farm incomes in the past have varied from year to year by up to 100 per cent, depending on the type of activity and the region. Income variability makes management a highly skilled task, and falls in income below that which are expected create financial and management problems.

Most farmers organise their business affairs to:

- have a good chance of the business surviving
- preserve the quality of the resources over the long term, given their implicit and explicit assessment of the chances of any one of the uncertain events happening, separately or in combination.

Farmers take actions on and off the farm in order to manage the risks they face. One of the best responses to production risks is to be very good at the animal and crop husbandry side of things. Often the best response to risks is to diversify the sources of **cash flow** to beyond the farm gate. It is common for farmers to have off-farm investments, and in some cases members of the farm family partnership earn some income off the farm.

In practice, farmers take many steps to establish themselves and their business in a position such that the risk of going bankrupt, or the risk of failing to meet other important objectives, is manageable for a wide range of circumstances. Whatever happens, their business has a good chance of providing the wherewithal to meet other important goals.

Farm systems

People often think that for a particular type of farm activity there has to be one best system for doing it. This is not so. In practice there are as many different systems as there are farmers. Over the years it has become evident that really good managers can make almost any system profitable, provided key requirements of size and control of resources are met. The reality is that really good managers are scarce. Only the best farmers are able to put all the bits of a system together well enough to make the system successful. The skill needed to get all the parts right to make the whole succeed is widely under-estimated and overlooked, particularly by people who take a partial view of the system.

The farm economy

The economic system that agricultural and horticultural businesses operate in is a market system. It is characterised by the following features:

1 most resources are owned by private individuals
2 production of goods and services is by privately owned businesses
3 decisions about what to produce and how much to produce are based on information coming out of free, competitive markets in which goods and services are sold and bought.

Producers buy the resources they use in markets. Doing this they decide how the nation's resources of land, labour and capital are used. Producers respond to what the buyers of goods and services are dictating in the product markets. If supply exceeds demand, prices paid to producers will fall. Resources will be re-allocated away from the 'unwanted' use and instead used for goods and services that consumers require. Competitive markets ensure that, of their many alternative uses, resources are used in ways that are most profitable.

Producers sell their products to consumers in markets. In markets the supply of products by producers meets the consumers' demand for products, and prices are formed. If demand exceeds supply at the going price, the price is bid up by competing buyers. If supply exceeds demand, prices are bid down. The central tenet of market-based economics is that competition between producers to acquire resources and convert them into the goods and services people need and want will force producers to do so as cheaply as possible (that is, to be economically efficient). If producers do not do this, then in the market where they sell their goods and services, the buyers will not buy the goods and services. They will get what they want from somewhere else. Competition among producers means that over time only those businesses that meet the buyers' needs and wants most efficiently will survive.

Competition to buy and sell products and services in markets is the essential characteristic of the economy of the Australian farmer. An individual farmer has to compete with every other farmer and every other type of business in the economy for the resources they need: the land, labour, fuel, chemicals, finance, transport services, risk-taking services such as insurance, machinery, feed supplies, animals. Having acquired the resources they need and produced their product, a farmer has to compete with other farmers in Australia and from overseas who are producing the same product and supplying the same market. Farmers have to compete successfully in the resource market to acquire the resources they need (that is, get them at a price at which they can make a profit). They then have to compete successfully in the product markets, by being able to sell their output at the going price and make a profit. If farmers cannot do this then their farm businesses will not remain profitable. The farmers will

FIGURE 3.1 The production and marketing chain

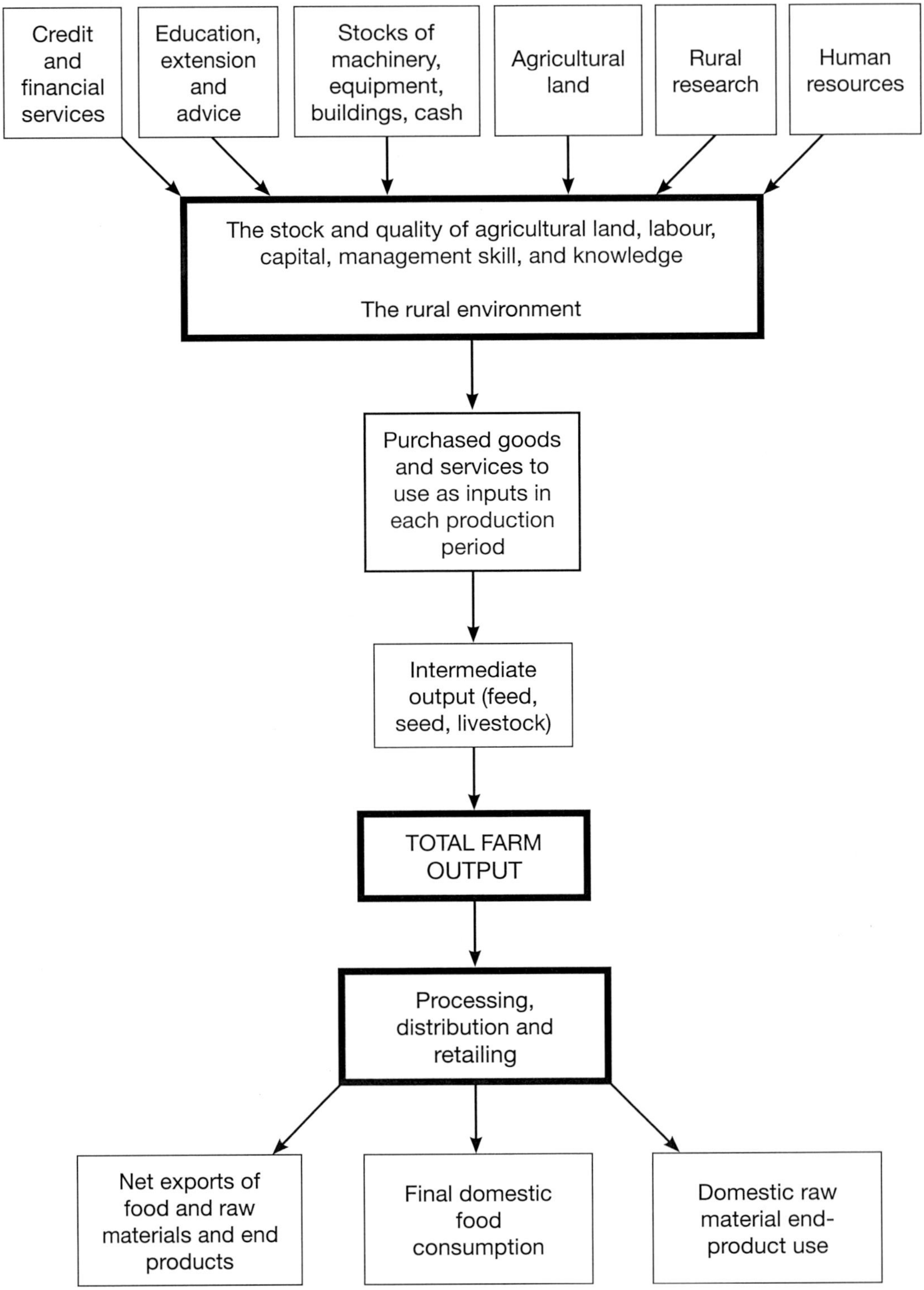

have to change what they do and how they do it. Change means moving into activities in which sufficient profits can be made. Or, change means selling their land, machinery and animals, moving out of farming and investing in some other business.

In market economies, the freedom to move out of one line of business and into another means there is no easy way to make profit. In the short term, an increase in demand for a product will cause prices to rise. This will make an activity profitable for those in it at the time. However, such pastures of plenty are short-lived. Other producers of other products, who can also produce the product that has become suddenly more profitable, will change their resource use into the more profitable form of production. In the medium term, supply of the profitable product will increase, relative to demand, and prices will fall and profits will return to the 'normal' level that existed before the boom. A prediction of economic theory is that in a market economy resources will continually move towards the most profitable uses, and the profitability of alternative uses of resources (or alternative forms of production) will tend to be similar over the longer term.

Conditions for success in Australian agriculture

The conditions for agricultural activities to succeed in Australia were evident early in Australia's history. To succeed the rule was: farmers had to be able to use large areas of land effectively with little labour, and their production had to sell profitably on export markets. This is still true. The relatively small domestic market for most agricultural products, and growing populations putting pressure on land elsewhere in the world, means that these conditions remain the keys to most forms of successful agriculture in Australia.

Davidson (1981) defined the conditions for successful agricultural production in a country with large areas of land and a small population, such as Australia, as follows:

1. agriculture must utilise a large area of land
2. agriculture must utilise little labour
3. agriculture must produce commodities that can be sold overseas
4. the commodity produced for export must be able to be transported long distances without losing quality. It must be valuable enough in terms of weight and volume to pay the cost of transport to the overseas markets.

These conditions were true for Australian agriculture from the beginning of settlement, and are all still true of broad-area agriculture. Conditions 1 and 2 are less applicable for intensive industries such as horticulture, viticulture and dairying, which have high inputs to land and, often, a high component of labour added to the product before it is consumed. Conditions 3 and 4 remain very much a fact of life for most agricultural activities.

Location

Why do agricultural activities occur where they do? The location of different farming is explained by climate, land topography and soil characteristics, distance from markets, and existence of related production activities both on and off the farm. The physical characteristics of soils, topography and climates found in particular places are determinants of the types of agricultural activities that can be undertaken. What is done on a farm is not only determined by what can be done in a technical sense, but ultimately by what is profitable with acceptable risk to the producer.

Technical conditions and transport costs are important determinants of profitability. A farmer, deciding what to produce, compares the profitability and riskiness of an activity in any particular region with profits and risks of feasible alternative activities in that region, and in other parts of the nation, and indeed in the rest of the world. The decision about 'what to produce' also depends on human factors. Farmers' goals, preferences, skills and attitudes to risk are all involved. The existence of complementary resources and activities are influential too.

Example of the role of transport costs on location of production

Suppose two products can be produced, product A and product B. Product A is bulky and perishable (e.g. fresh milk). It is costly to store and transport. Product B is able to be stored and is cheap to transport. Product A is more profitable than product B for the farmer to produce if the farmer receives at least a net 20c/litre for product after transporting it to market. If product A sells for 30c/litre in the market and costs 1c/litre/km to transport, then it will be produced within 10 km of the market. Beyond 10 km, the transport cost reduces the return to the farmer below the 20c/litre at which the profit from product A breaks even with the profit of product B. Beyond 10 km from the market farmers will produce the more profitable alternative, product B. Thus transport costs help to determine 'what happens where' by setting the boundaries or limits of production.

The effect of transport costs on location of production is that products that have high per unit transport costs relative to the prices they sell for in markets will be produced closer to final markets than other products with lower transport costs. For example, before transport and storage costs reduced markedly over time, perishable bulky fresh milk or fruit and vegetables that were costly to transport and store (e.g. requiring refrigeration) relative to their market price were produced close to where they were sold. Production of storable grains and meat that were less expensive (e.g. rail) to transport relative to their market price were able to profitably locate further away from their end markets.

Modern storage and transport has reduced the costs of transporting agricultural and horticultural products and extended the boundaries of production. All forms of agricultural production are able to locate in places that were not possible in previous times, though the cost of transport in determining profitable locations continues to play a dominant role in determining profits and thus determining what happens where.

Marketing chain

The various input and output stages involved in the production and marketing of agricultural commodities are shown in figure 3.2. Farmers link forward to the consumer through the marketing chain and backward to the suppliers of farm inputs. Farmers buy products and services from producers of other goods and services in the economy, such as supplies of fertiliser, machinery, chemicals, medicines, seed, breeding stock, insurance, credit, transport, fuel, equipment and so on. Farmers supply food and fibre products for consumers. Also, farmers supply inputs to other farmers.

Mostly farmers sell to middle-people rather than directly to the consumer. Other businesses transform the farm product. Farm products have to be transported, processed, financed, distributed, and sold. Farm products are produced in batches, not continuously. To convert intermittent production into a continuous stream, storage is needed. Holding stocks involves the risk that prices will fall in the interim. Specialised risk takers, speculators, are important in marketing farm products. The complexity of agricultural marketing increases because many farm products are bulky and perishable. An efficient market channel enables consumer needs and wants to be met as cheaply as possible. Efficient **marketing systems** respond rapidly to changes in what consumers want and how they want it.

Marketing is all the processes that are necessary to transform a farm product into something consumers want, when, where and how they want it. Information is a key element in these processes. Market prices inform producers, consumers and everyone else involved in the marketing chain, from input suppliers to processors, to financiers and risk takers. Information makes markets work better, in terms of prices reflecting better the characteristics of the product (called pricing efficiency) and in facilitating more efficient handling, transport and storage of products (called operational efficiency).

Farmers enter into strategic alliances with other businesses along the marketing chain to achieve efficiency gains, by improving information flows, improving product quality or consistency, and earning price premiums for providing such improvements.

FIGURE 3.2 The markets

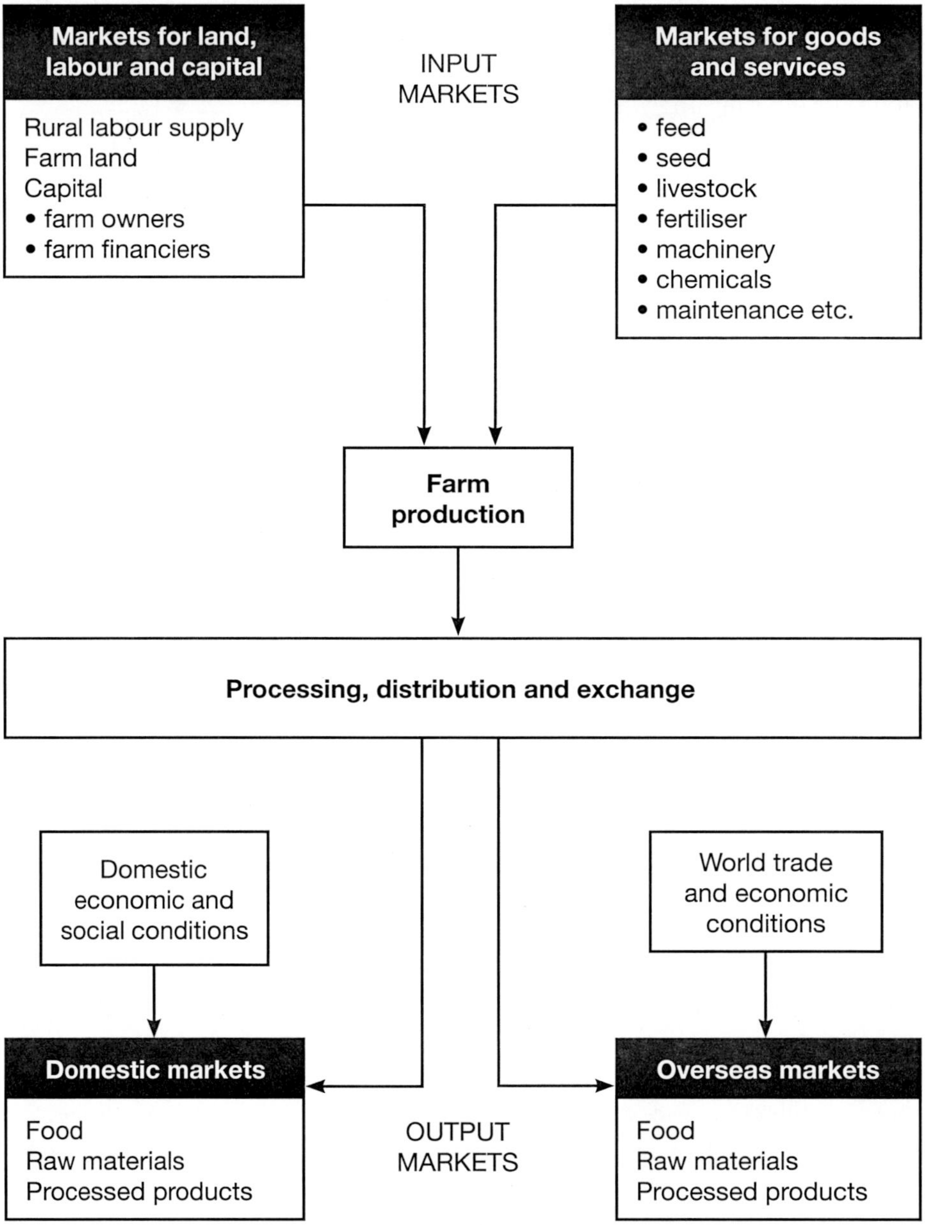

Agriculture and economic growth

Agricultural output from farms has always been an important part of the economic growth of countries. Economic growth means greater income and goods and services available per head of population; people are better off; and they have greater control over the environment in which they live. The increase in the capacity of an economy to produce goods and services comes from advancing technology and changing ways of thinking about production. Such changes are continually required for economic growth. Nowhere is this truer than in the agricultural sector striving to provide food and fibre, export earnings, labour for industrialisation, capital for investment and environmental services too.

Just as agriculture affects economic growth, economic growth affects agriculture. Economic growth causes the following changes to occur:

- the proportion of the workforce in agriculture declines
- agricultural output declines as a proportion of the total market value of all goods and services available in a country in a year (as the rest of the economy expands more rapidly)
- agricultural output per labour unit remaining in agriculture increases greatly
- the non-agricultural sector of the population places higher value on environmental goods and services, many of which are controlled, managed or affected by farmers.

In these processes the relative returns to capital and labour invested in the non-agricultural sector and in agriculture are key determinants of the rate at which these changes occur.

Although the size and significance of the role of agriculture in the Australian economy has changed over time, agriculture produces more gross value of product and value of exports than ever. The role of agriculture in the economy has declined as the role of other sectors has increased. Agriculture contributed 60 per cent of the gross domestic product in 1900, 20 per cent in the early 1950s, 14 per cent in the early 1960s, 7 per cent in the early 1970s, 5 per cent in the early 1980s, 3 per cent in the early 1990s and 2 per cent in the early 2000s. Agriculture supplied 90 per cent of total exports in the early 1950s and 15 per cent in the 2000s. In 1961–62 rural employment was 11 per cent of total Australian employment. In 1986–87 this was 5 per cent, and in 1994 it was 4 per cent. In 2007 it was 2.4 per cent. Total full-time employment in agriculture in the mid 1960s was about 420,000. In the early 2000s total full-time employment in agriculture was around 375,000 people. More output from fewer people means productivity of labour has increased.

Basic characteristics of the farm economy are:

- demand for farm products responds little to changes in price
- total demand for farm products responds little to changes in income
- there is a competitive structure
- rapid technological change occurs
- farm resources such as land, labour and machinery and equipment do not move easily out of agriculture into other uses in the economy.

Economic growth causes income in the non-agricultural sector to rise faster than in the agricultural sector. This phenomenon arises partly because demand for agricultural products does not respond a great deal to falls in prices, or to rises in incomes of consumers. Hence as an economy grows and national income increases, the proportion of extra income spent by people on consumption of food declines. Inevitably the share of national income going to agriculture declines relative to the share going to the rest of the economy.

As well, supply of agricultural commodities increases more rapidly than demand because of technological change, prices of farm products fall, and farmers have to increase output to maintain income. They do this by adopting new technology, and the cycle goes on. Without a strong export market to absorb the increases in supply, farmers are trapped on a so-called 'treadmill': they have to produce more (run faster) to maintain incomes similar to people in the rest of the growing economy.

The problem of falling prices and rising costs putting pressure on farmers to improve their productivity is exacerbated because farmers' assets, including themselves, are often in their best use, despite earning low returns. The extent to which a problem of lower per capita incomes in agriculture than in the rest of the economy exists is partly determined by the rate farmers move out of farming. This in turn depends partly on the extent to which their assets have profitable alternative uses and on the employment opportunities and wages in the non-agricultural sector. In this way, the non-agricultural sector influences the structure of agriculture.

As well as the above reasons, an important way economic growth affects the farming sector is through competition between the agricultural and non-agricultural sectors of the economy for capital and labour, and for services such as transport as an economy grows. Farmers have to compete with the non-agricultural sector for their inputs. If the non-agricultural sector is growing faster than the agricultural sector then non-agricultural businesses can afford to pay more for resources such as labour, capital and land, than people running farm businesses can afford. The overall effect is that problems caused by declining real prices received by farmers are made worse by rising real costs of production. Farm incomes are further 'squeezed'.

The 'squeeze' on farm incomes caused by declining prices received per unit of output, and increases in costs per unit of output, does not necessarily mean declining incomes for farmers. Most farmers offset the effect of this cost-price squeeze on net farm income by increases in farm productivity.

The economic forces farmers confront leads to the effect over time that economic growth causes either fewer farmers (who earn an income regarded by the farmers as sufficiently similar to other uses of capital in the economy, once non-economic considerations are considered), or poorer farmers. The third option is for farmers to be subsidised by taxpayers and consumers. This option has been widely adopted in the USA and in Western Europe, and is inefficient and inequitable. The general explanation of economic growth and its effects on agriculture predicts a decline in the relative share of national income, and employment, and a decline in the number of farmers over time. This has occurred in Australia.

Change in the farm economy

Farming is always changing. The characteristics of the human, technical, economic, financial, risk and institutional elements that comprise farm businesses and combine to create farm products are always changing. Every harvest cheque received on every farm in every year is a result of different combinations of technologies and inputs, and different natural and economic conditions. Rapidly, prices and exchange rates change, types of farm activities appear and disappear, methods of production change; technology becomes outdated and new technology is adopted; diseases break out; consumer incomes and tastes change; markets grow and diminish; competitors come and go, strengthen and weaken; laws change; farmers change. Each production and marketing situation that a farmer manages is different to any previous situation.

The natural conditions in which agricultural product is to be grown and the economic conditions that will prevail when it is to be sold cannot be known at the time the farm production decisions are made. Decision-makers form expectations about the future, make a decision, act on it and then respond as the future arrives and turns out different from that expected. The public too faces the same situation. Governments have at times implemented policies to do with agriculture to achieve certain aims, and subsequently had to implement new policies to achieve the opposite aim because situations changed. For example, policies to establish large numbers of small farmers were followed by public policies to help small farmers leave agriculture; subsidies for fruit production were followed by subsidies to pull out trees.

Change comes to farmers in the form of new information. The future cannot be known, so the best that can be done when running a farm business or implementing farm policy is to make good decisions based on the best information available at the time. When the future turns out differently from the expected, that is, when change occurs, farm business-people and policy-makers renew expectations and act in response to this new information. One criterion to judge how change will be accommodated by farm businesses is the extent to which the change is a surprise. If a particular change is able to be anticipated with some degree of probability (called risk), then it is less surprising than something that cannot be expected (uncertainty).

In the litany of change in agriculture described below, some changes such as drought and short-term fluctuations in prices within a reasonable range are known elements of the farming game, though their timing is unpredictable. Other change, such as long-term market collapses or loss of access to markets because of disease, or changes in consumer preferences, or emergence of new competitors, or invention of new inputs and processes, are less predictable and more surprising to farmers and policy-makers.

The nature of changes and pace of changes in agriculture has to be seen in the context of their times. Farmers in the future will not have more nor less profound change to manage, in the context of their times, than the changes their forerunner farmers have had to deal with in their past. The future changes that future farmers will have to manage will not be faster or slower, in their future context, than the rates of change that farmers have confronted at different times in history.

Changes that farmers in Australia have experienced since the first European settlers arrived include:

- price rises and falls
- costs rising and falling
- changing exchange and interest rates
- loss of access to markets
- changes in demand for products
- changes in supply by competitors
- provision and removal of public assistance of various forms
- changes in land law and tax law
- unexpected deficiencies of rainfall, sometimes for long periods
- new technologies, machinery, equipment
- computers
- outbreaks of pests and diseases, control of pests and diseases
- growth of other sectors of the economy
- competition from other sectors of the economy for resources
- introduction of subsidies and protection of agriculture and non-agriculture that disadvantages export-oriented agriculture.

In the beginning, the first pastoralists moved from coastal settlements and ran the well-favoured inland areas as a large sheep run. The 1830s brought the first wool boom. This was followed quickly by an economic downturn that reduced the worth of sheep to the value of boiled down tallow. By the 1840s, agriculture had experienced a major drought, and its first boom and its first bust: events that have continued to characterise Australian agriculture. Discovery of mining and export of gold in the 1850s and 1860s kicked off a sustained period of strong economic activity that lasted until the 1890s, albeit with serious droughts in each decade.

From 1850 to 1900 the Australian economy had its first long economic boom. During the long boom there were massive inflows of capital and labour; financial institutions and capital markets grew; there were runs of good seasons; land legislation was enacted and settlement took place; numerous technical advances were made. Major technological developments in this period were in road and rail transport, scrub rollers, stump jump ploughs, stripper harvesters, fencing and water delivery.

The geographic frontiers of crop and livestock production progressed; rabbits proliferated; venture capital investors lost as plantation-style investments in cotton and sugar in north Queensland failed; an epidemic of Redwater fever devastated the cattle industry. There were periodic severe droughts; commodity and land markets rose; there was much speculative investment in land; huge unpayable debts were incurred and, in the 1890s, a **liquidity** crisis in British financial markets caused bank collapses and financial havoc in the colonies.

The wool industry grew rapidly: based on cheap land, development of the Merino sheep and low labour requirements, aided by strong demand from large markets in Europe. Australia's sheep numbers grew from 20 million in 1860 to 100 million by the 1890s, falling quickly to 50 million as the prolonged Federation drought from 1895 to 1902 wiped out sheep feed.

Dairying developed close to all the capital cities where there was high rainfall and good soils, helped by the development of railways, refrigeration, milking machines, food science and factory processing of butter. Railways and the new input phosphorus brought inland regions into cereal production. In the 1880s irrigation schemes commenced along the major rivers.

By the start of the twentieth century the institutional features that were to determine the shape of agriculture in Australia were already established. There was government-sponsored agricultural research and extension; closer settlement of agricultural landholdings; public irrigation schemes; protection for agricultural and other industries. The future was evident: agricultural activities experiencing good and bad seasons and prices, while the rest of the economy grows rapidly and simultaneously competes with agriculture for scarce land, labour and capital while adding to demand for farm output.

Closer settlement of land by small farmers

Schemes for closer settlement of land by small farmers were implemented following the First and Second World Wars. The pattern of agricultural activities that exists across the countryside today has been heavily influenced by public land settlement schemes. Australia implemented closer settlement schemes from the 1860s, and following the two world wars of the twentieth century. The rationale for Closer Settlement, fallacious as it turned out, was that agriculture could provide a living for a large number of small farmers. The belief was this would develop the countryside and stimulate wider economic growth. Large landholdings were subdivided into small landholdings. Efforts by governments to settle the land with small farmers continued until the 1970s.

The first Soldier Settlement Scheme was implemented after the First World War, settling returned soldiers on the land. Some settled on newly opened Crown land; the rest on parts of large pastoral properties. Small dairy farms were established, market gardens set up close to large centres, and ex-soldiers were sent to sheep properties in Victoria and on wheat farms in marginal agricultural areas.

In the irrigation districts, blocks were planted to citrus and vines. Although all of these activities were already occurring, or since then have successfully been carried out in these areas by private entrepreneurs, many of the soldier settlers were engaged in a battle they could not win. With little capital, battling for a living off small, rough blocks of farmland, many were armed inadequately to fight the market collapses and economic perils of the Great Depression of the 1930s.

Further Closer Settlement for ex-servicemen followed the Second World War. This time some training was given to the novice farmers, and generous and more flexible financial arrangements were provided. A lesser, twentieth-century version of the previous century's long boom occurred during the 1950s and 1960s, enabling most settlers to do reasonably well until the downturn in farming fortunes in the late 1960s. This exposed the vulnerability of small farmers to the effects of declining prices for their products and rising costs of the inputs they required.

The biggest single reason for the failures of Closer Settlement was inadequate farm size. Inadequate capital and management ability were also common problems, particularly with the earlier schemes. The capacity of any particular sized farm business to generate a living depends upon the physical characteristics and the ruling economic circumstances, the run of the seasons and the proportion of the farmer's capital and debt. 'Adequate' farm business size is ever-increasing.

Closer Settlement was replaced in the 1970s by various schemes aimed principally at helping excess farmers to leave the land and, in an about turn, helping farmers whose farm was too small to expand: 'get big or get out or become reconciled to low incomes' was the message of these rural adjustment schemes.

Innovation and change

Agricultural **innovation** continued apace into the twentieth century. The motivation for many innovations was to help farmers manage the problems of farming soils with low fertility under conditions of low and highly variable seasonal rainfall. Wheat was bred with growing requirements that matched short growing seasons and low rainfall. By the 1920s phosphatic fertilisers introduced for cereals were adopted for pastures. Motor vehicles arrived and tractors with rubber tyres replaced steel-wheels and stationary engines. Implement sizes started increasing in size, and never stopped.

In the 1930s the Great Depression caused massive unsaleable surpluses of farm products and low prices. Low farm incomes and widespread hardship prevailed. Farm businesses on Closer Settlement schemes set up after the First World War failed.

The Second World War brought economic recovery and regulation of farm prices, with serious widespread drought in 1943–45. The second long economic boom that commenced following the Second World War continued until the mid 1960s. There were periodic serious dry spells in each state. The 1950s was a time of significant public investment in rural infrastructure and rural enterprises. Land was cleared: the area of Australia's farmland increased significantly. Irrigation dams and schemes were built. Transport improved. Ley farming and fallowing to conserve soil moisture for crops evolved. Bulk handling replaced bagged grain. Fertiliser and nitrogen-adding leguminous pasture plants led a grazing revolution in the 1950s that doubled carrying capacity of pastures. Correction of trace element deficiencies in soils expanded croplands and yields, and increased livestock productivity. New and highly effective pesticides and herbicides, fungicides and animal health treatments became available regularly. Power sources and hydraulics found myriad uses on farms. Farm labour productivity rose to hitherto unimaginable levels.

Wool prices went to record high levels in 1950. Australia experienced a balance of payments crisis in 1950/51. This was the time of the Fixed Exchange Rate Agreement. This gave agriculture the central role in earning export income and balancing the nation's Balance of Payments. Agriculture had to earn the export income to buy the imports that were required for the development that was taking place. Consequently agriculture was protected in various ways with regulations to protect farmers from competition, subsidies on inputs and outputs, price schemes to raise returns to farmers, special tax concessions and credit arrangements for farmers.

By the late 1960s and 1970s, the good times in agriculture were passing. The fixed exchange rate regime ended, and the British pound devalued. In 1962 the United Kingdom had joined the European Common Market. In 1971, Australia's preferential trade access to the United Kingdom market ended. Exports of Australian beef and butter to the United Kingdom fell from 80–90 per cent in the 1950s to 10 per cent by 1975. Through this time protection of farmers in the European Economic Community and Japan increased markedly. Access of Australian produce to these markets was restricted severely. At the same time the USA increased assistance to farmers. World supplies of agricultural products grew and agricultural prices trended down.

Wheat, wool and dairy surpluses in the late 1960s brought the introduction of quotas in wheat, subsidies added to low wool prices and eventually a reserve price scheme, and poverty in small sized dairying. Serious welfare and efficiency problems of small sized dairy, wheat and wool farms led to the re-emergence in the public domain of policies to help farmers cope with change—rural reconstruction, later rural adjustment schemes with the aim of speeding up the rate of change of small unviable farmers out of agriculture.

By the 1970s, the secondary and mining industries had emerged as major sectors of the economy. Mining earned more export income than agriculture. The mineral boom raised the exchange rate above the level it would otherwise have been. This adversely affected agriculture. In 1973 world oil prices quadrupled, and **inflation** rose to record high levels. In a sign of future changes in the economy, tariff protection to Australian industries was cut 25 per cent. Through the 1970s, wheat, wool and dairying suffered periodic crises caused by the difficult double of losing access to export markets and world oversupply. When not in crisis, there were also some profitable times. On-farm, conservation farming techniques started to be adopted, wool was measured objectively, and intensive egg, broiler and pig production changed rapidly from small sideline operations to integrated contract farming operations.

The second oil price shock occurred at the start of the 1980s, prompting recession. International farm protectionist policies caused falling farm prices and incomes for exporting countries like Australia. Protection of domestic secondary industry adversely affected the exchange rate and disadvantaged agriculture. The imperative for increasing efficiency in Australian agriculture and the economy at large continued to grow. Australia pursued policies of reducing protection to all industries, including agriculture. Another serious and widespread drought occurred in 1982. In 1983, the exchange rate was floated, introducing further volatility for farmers to manage. The financial system was deregulated in 1983. Financial deregulation changed the supply and conditions of credit and volatility of interest rates, and increased the sophistication of financial products, lending and risk assessment. In 1987, a **capital gains** tax on assets was introduced.

By the mid-1980s wool prices under the wool reserve price scheme had been set at unrealistically high levels. Record sheep numbers built up, with record production of wool. Demand for wool fell. The Australian Wool Corporation bought wool to keep the price high, and built up a massive stockpile of wool. This scheme collapsed in 1991, leaving woolgrowers and taxpayers seriously out of pocket. There has been a free market in wool ever since. The Australian wool industry has diminished in size, with significant diversification into prime lamb production.

The third major period of strong economic growth occurred from the mid 1990s to 2007. The flow of technological changes continued. The domestic wheat market was deregulated in 1989. Egg production was deregulated and quotas removed. Dairy production was deregulated in 2000. These changes each had significant effects on how farmers in these industries operated. In the 1990s farmers in all industries around Australia experienced serious drought and price declines, plus the odd boom in prices and good seasons. Outbreaks of disease in the United States beef industry benefited Australian beef farmers by enabling increased sales to Japan, which had banned United States beef. Changes in tax laws encouraged new investment in agricultural activities. Sophisticated price-risk management options that enable farmers to sell some of their price risk to people willing to bear that risk were used more.

Nominal value of rural land in most parts of Australia grew strongly from 2000. In 2003, dairy farmers ran into the triple impacts of drought and reduced irrigation water, high feed and water costs and low milk prices. All dairy farmers incurred significant extra debt and many moved out of the industry. In 2006–07 more serious drought and water shortages prevailed. Irrigation capacity was reduced and feed costs increased. More debt was incurred and more dairy farmers moved out of dairying.

In 2003 and 2006–08 severe and widespread droughts reminded new generations of farmers of the fragility and the vulnerability of their livelihood. As has always occurred with droughts, governments responded with a series of measures to ameliorate the effects on the farming community. Drought assistance measures penalise those farmers who prudently plan for the inevitable drought.

Oil prices doubled in the period 2005–08, then fell. Technological change continued to rapidly create opportunities, changing what farmers do and how they do it. Computerised operation of machinery and equipment and management of information brought productivity progress in a wide variety of ways in all types of farming, from precision agriculture in cropping operations to animal management in both intensive and extensive livestock production to plant and livestock genetic improvement. New varieties of plants better adapted to natural environmental conditions are becoming available rapidly. Opportunities for closer relationships throughout marketing chains are growing. Farm sizes and output per labour unit have continued to grow steadily. Real prices have continued to fluctuate around a declining trend and real costs keep on increasing. In 2007, real wheat prices doubled as harvests halved, and costs rose significantly. The best cure for high prices is high prices and in 2008 supply increased and commodity prices fell from their high levels back towards long term average levels. In late 2008, a global financial crisis caused economies all around the world to move into recession. Demand declined, exchange rates fell, while government spending increased and interest rates were reduced in attempts to simulate economic activity—all part of a days work in farming.

The environment in which Australian agribusiness firms operate has changed markedly over the past three decades. Deregulation of the exchange rate, the financial sector, labour markets, and liberalisation of trade arrangements have led to greater exposure to competition in all markets. A major consequence of increasing deregulation of economies is greater uncertainty. At the same time, more avenues to manage volatility become available. Further, changes in the level of economic activity and in key economic variables like rates of foreign currency exchange, or interest and inflation, have more immediate and direct impact on the welfare of individual farmers than was the case in the more regulated business environment of the past. International economic developments also directly affect individual economies to a greater extent than in the past. For instance, flows of capital are more readily 'internationalised' than ever before. Nowadays the effects of changes in the economies of the major developed countries, as well as in the lesser developed but developing countries, are likely to translate more readily into effects on Australian businesses, including farmers.

The capacity of management to manage a business is really the capacity to manage change. As ever, farmers have to manage the imperative of increasing productivity to maintain real returns on capital. The challenge of managing farm businesses in an environment of frequent and significant change has two profound effects:

- change creates the need for constant increases in agricultural productivity
- change also provides the means of achieving the necessary increases in agricultural productivity.

As it has happened the record of many of farmers in Australia at increasing productivity is better than for most other business-people in other industries in the economy.

REVIEW QUESTIONS

1 What are common goals of farm families?
2 What returns to capital are earned by large, medium and small farm businesses?
3 As a nation grows wealthier the share of national income contributed by agriculture declines. Why?
4 The nature of supply and demand of agricultural products has major implications for farming. What are the main characteristics of agricultural supply and demand?
5 Real prices received for farm products decline over time, and real costs of producing agricultural products increase. Why does this happen, and how do farmers respond to remain profitable?
6 With economic growth, fewer farmers or poorer farmers or highly subsidised farmers is the rule. Why is this so?
7 Coping with change is the essence of farm life. What are some of the main changes farmers in Australia have dealt with in the past?

FURTHER READING

Davidson, B. R., *European Farming in Australia*, Elsevier, Amsterdam, 1981.

Reid, R. L. (ed.), *The Manual of Australian Agriculture*, Butterworths, Sydney, 1990.

WEBSITES

Australasian Agribusiness Review and Australasian Agribusiness Perspectives: www.agrifood.info/

Australasian Farm Business Management Network: www.csu.edu.au/faculty/science/saws/afbmnetwork/

Australian Agricultural and Resource Economics Society: www.aares.info/

Australian Bureau of Agricultural and Resource Economics: www.abareconomics.com/

Australian Farm Institute: www.farminstitute.org.au/

CSIRO: www.csiro.au/

National Farmers Federation: www.nff.org.au/

NSW Department of Primary Industry: www.dpi.nsw.gov.au/

Vic Department of Primary Industry: www.dpi.vic.gov.au/dpi/index.htm

PART 3

Fundamental Processes of Plant Production

Principles of Plant Production

4

CONTENTS

Introduction

Crop and pasture plants are special within the plant world. These plants are grown in 'managed' populations in order to produce food and fibre products for humans. There are more than 230,000 species of flowering plants in existence, classified into some 315 plant families. People have selected and modified only a small number of plants, from just a few families, to use in agriculture. Only 15 plant species provide the great bulk (more than 80 per cent) of the food consumed by humans. More than half of the human food worldwide is provided by eight species (wheat, rice, maize, barley, sorghum, millet, oats, and rye). These are all from the same grass family: the Poaceae. Two plant families contain most of the important plants used in Australian agriculture. These are the Poaceae and the Fabaceae (legumes).

Farming in Australia involves managing communities of crop and pasture plants so that they grow abundantly and efficiently. Often the environment these plants grow in is less than optimal because of variable rainfall or infertile soils. To appreciate how to manage plant populations, it is necessary to understand plant growth. This chapter is about the fundamental processes of plant growth, and how plants are influenced by environmental conditions.

How crop and pasture plants grow

Plants need to acquire five or six basic resources from the environment in order to grow. These resources are available in the biosphere. They are:

1 solar radiation, which provides light and favourable temperature regimes
2 carbon dioxide (CO_2) from the earth's atmosphere
3 water from the soil
4 mineral nutrient ions that are dissolved in the soil water
5 oxygen (O_2) from the soil atmosphere
6 and nitrogen (N_2) from the soil atmosphere (this is required by legume plants).

Plant productivity on farms depends on sufficient amounts of these resources being available for crop and pasture plants so that they can grow abundantly and accumulate dry matter. Light energy from the sun instigates a series of chemical reactions called photosynthesis. This process uses carbon dioxide and water to produce the organic molecules that form the plant tissues. The atmospheric gases are abundant and non-limiting. Soil water on farms depends on rainfall unless irrigation water is available. The amount of water stored in the soil for plant use can to some extent be manipulated by the farmer, as can soil aeration, which aids the supply of oxygen to the plant roots. The supply of mineral nutrients in the soil can be increased by the use of fertilisers, and the supply of nitrogen in the soil can be increased by growing legumes.

Structure and function of plants

Plants have developed special structures that enable them to acquire the environmental resources that are essential for their growth. A branching root system that grows through the soil gathers soil water and dissolved nutrient ions. The root system attaches to the base of a plant stem at the soil surface and provides a firm anchorage into the soil. The stem in turn provides the physical support for the leaves, which are positioned above the soil. The leaves are attached to the stem or stem branches by petioles or leaf sheaths at small ridges on the stem called nodes. The leaves contain chlorophyll pigments that absorb incoming solar radiation. A waxy cuticle surrounds plant leaves, which prevents the plant losing moisture to the atmosphere. The leaf surface contains special cell pairs known as stomatal guard cells, which form openings in the leaf surface to allow atmospheric CO_2 to diffuse into the tissues within the leaf. Specialised cells connect to form strands of conducting tissue, called xylem and phloem, which extend from behind the root tips to all parts of the leaf canopy. The xylem carries water and dissolved nutrients upwards from the roots to the leaves, while the phloem transports sugars and other metabolites from the leaves to other plant parts.

An essential feature of all plants is their ability to reproduce. Reproduction occurs following a period of vegetative growth and culminates in the formation of flowers, arranged singly or in groupings called inflorescences. These are located in the upper part of the shoot system in most plants. The flowers develop into seed-containing fruit when the plant matures. The successful production of harvestable seed is the objective of growing crop and some pasture plants. The seed fraction is generally the harvestable portion for crop plants, so the task of the farmer is to produce plenty of seed (or grain) from the crop. In the case of pasture plants the seed is required to establish new pasture plants in the soil.

Flowers of plants have a basic plan consisting of a series of plant parts arranged in circles or whorls around a central axis. The bottom whorl commonly consists of green, leaf-like structures called sepals (collectively the *calyx*), which protect the developing flower bud. The next whorl consists of the conspicuous petals (collectively known as the *corolla*). These may be highly coloured (especially in the case of insect-pollinated plants). The male reproductive structures, or *stamens*, are located within the petals; each stamen consists of a slender filament supporting a pollen-bearing anther. The central part of the flower is a structure called a *pistil* containing one or more of the female portions of the flower, called a *carpel*. Each carpel consists of an enlarged base, or *ovary*. Carpels contain ovules. The base extends upwards to form an elongated style that has an expanded tip, or *stigma*. The stigma is where pollen is deposited, and the style is the means of delivering germinated grains of pollen, called male nuclei, to the flower's ovaries. Pollen grains from the flower's anther (in self-pollinated plants) or from another plant (in cross-pollinated plants) are transferred to the stigma by insects, birds or the wind: when a grain germinates on the stigma, it forms a pollen tube that carries nuclei down through the style to ovules in the ovary. Fertilisation of the flower occurs when the male and female nuclei fuse in the ovary. The fertilised ovule develops into the seed while the mature ovary becomes the fruit.

The reproductive structures of flowering plants provide distinguishing characteristics used in plant identification. Agricultural plants in Australia come from a number of plant families, the most important being the grass family, the Poaceae, and the legume family, the Fabaceae. There are major differences in the form and appearance of grass and legume flowers, as the two groups of plants use different agents to deliver pollen to their stigmas. Many legumes are 'open-pollinated' and rely on insects to deliver the pollen. For this the legume flower has a specialised five-petal arrangement and nectar-producing nectaries at the base of the staminal column to attract insects. When an insect alights on the horizontal wing petal, the bottom keel petal is 'tripped' and this sensitises the stigma to be receptive to the pollen deposited by the insect. The grasses have developed a much less demanding pollination process that relies on the wind to deliver pollen to their stigmas. The grass flower or floret is a simple structure with no petals, an ovary containing one ovule, two styles with feathery stigmas, and usually three stamens. These are enclosed by two enclosing leaf-like structures, called the *palea* and the *lemma*. At flowering, the lemma and the palea are forced apart by the swelling of two small components at the base of the ovary, called *lodicules*. Rapid elongation of the slender stem (the *filament*), which has the pollen-producing anthers attached to it, causes the anthers to protrude and release pollen for wind dispersal.

FIGURE 4.1 (a) The inflorescence, and (b) the five-petal flower of the legume *Genista linefolia*; (c) the inflorescence, (d) spikelets or groups of florets, and (e) an individual floret of the annual grass *Poa annua*

Growth and development

Growth is the increase in the size and weight of the developing plant. The accumulating dry matter associated with growth consists of organic compounds such as cellulose, starch, lipids and proteins. Growth is brought about by cell division and cell expansion. Cells divide in special regions called meristems.

These are located in different parts of the plant. Increase in stem growth results from the expansion of cells produced by cell division in the apical meristem, located just behind or beneath the stem tip. Leaf growth in grasses results from cells produced by cell division expanding in the interscalary meristem at the leaf base at ground level. New leaves are pushed upwards from the protected intercalary meristem—a feature that makes grasses well adapted to being defoliated by grazing animals. Similarly, root extension through the soil results from the continual division of cells in the meristem of the root tip, followed by their expansion and differentiation into root tissue. Plant development is the passage of the plant through its life cycle and involves a sequence of phases. The phases for crop plants involve crop establishment, then vegetative growth, followed by reproductive growth that culminates in seed production. Annual plants then senesce and die, whereas perennials experience a resurgence of vegetative growth after seed formation.

Processes involved in plant growth

Growth processes in plants involve biochemical transformations of carbon-containing organic molecules in plant cells. They invariably involve synthesising new organic molecules or breaking down existing molecules. These reactions require a source of biochemical energy and a supply of electrons from 'electron-supplying' (or reducing) compounds. Energy for plant cells is supplied predominantly in a co-enzyme molecule called adenosine triphosphate (ATP). This is the biochemical 'fuel' or energy source for the plant. The supply of electrons is provided by the reduced forms of the compounds nicotinamide adenine dinucleotide (NADH) or nicotinamide adenine dinucleotide phosphate (NADPH).

Photosynthesis

Photosynthesis is fundamental to the growth of plants. It involves two separate processes. The first process requires sunlight and converts light energy from the sun into chemical energy. This occurs within structures in the leaf cells called chloroplasts. This conversion is caused by the flow of electrons from water molecules, through pigments (principally chlorophylls) in the chloroplasts. The electron flow results in the reduction of the co-enzyme NADP to NADPH and the addition of a phosphate ion to the co-enzyme adenosine diphosphate (ADP) to form ATP. Water molecules are broken down in this process, with the hydrogen atoms becoming part of the NADPH and the oxygen being released in the chloroplast. The result is that energy in light is transformed by the chloroplast pigments into chemical energy, which is contained in reductant NADPH and the energy-rich ATP molecules.

The second process in photosynthesis also occurs in the chloroplasts of leaf cells. It provides the plant with carbon compounds for growth. In this process (called the C3 pathway) CO_2 is assimilated with water and a 5-carbon compound, ribulose biphosphate (RuBP), to form two 3-carbon phosphoglycerate molecules. The atmospheric CO_2 diffuses into the sub-stomatal cavities. These are close to the chloroplasts in the adjacent leaf cells; water is delivered to the leaf cells from the roots via the xylem conducting tissue; and RuBP is synthesised within the chloroplast. The catalyst for this process is an enzyme, ribulose biphosphate carboxylase. Subsequent reactions use the products of the light reaction, NADPH and ATP, to convert the phosphoglycerate molecules into other 3-carbon compounds, and then combine these into one 6-carbon sugar compound (fructose 1,6-biphosphate). This sugar is the precursor of carbohydrate compounds (compounds consisting of carbon, hydrogen and oxygen). Carbohydrates include sucrose, starch and cellulose. Plants use these compounds for metabolism and growth. The sugar can also be broken down to provide energy for metabolic processes.

The reactions involved in photosynthesis proceed rapidly with plenty of light and adequate temperatures. In plants adapted to temperate conditions, the result is that there is a build-up in oxygen within the chloroplasts, and a decline in the CO_2 levels. There is also a build-up in reducing compounds that can damage the cells. These conditions lead to 'photorespiration', which consumes reducing compounds. In this process, oxygen (instead of CO_2) combines with RuBP, leading to a loss of RuBP. Plants minimise this loss of RuBP by processing reaction products of photorespiration back to RuBP. Energy is expended and some carbon is lost as CO_2.

The C4 pathway

Crop and pasture plants can be divided into two groups according to the way that they incorporate CO_2 into organic compounds. Most temperate crop and pasture plants incorporate CO_2 into RuBP to form two 3-carbon molecules of phosphoglycerate using the C3 pathway already described. At high temperatures (>30°C) these C3 plants have high rates of photorespiration. This limits the extent to which they can use the high light energy for further growth; that is, they are unable to increase photosynthetic rates and growth rates. The second group of plants are C4 plants, and includes many plants that grow in tropical areas. These C4 plants assimilate CO_2 in a way that minimises photorespiration. This enables better use of the abundant light energy, and thus high rates of growth are possible. The C4 plants incorporate CO_2 into a 3-carbon molecule (phosphoenolpyruvate or PEP) to form 4-carbon organic acids. The acids are formed in the chloroplasts of mesophyll leaf cells, and are then translocated into the chloroplasts located within adjacent bundle sheath cells that surround strands of xylem and phloem tissue. Here the CO_2 is released. Subsequently the CO_2 is incorporated into RuBP via the C3 pathway. The C4 photosynthetic mechanism allows plants to use CO_2 and water more efficiently, especially when light levels and temperatures are high, than is the case with the C3 plants. The strong affinity of the 3-carbon molecule PEP for CO_2 in the C4 plant leaf means the plant is able to maintain low CO_2 concentrations in the sub-stomatal cavities within the leaf. This means there is a steeper CO_2 gradient from the atmosphere into the leaf tissue, even when the stomata close partially in response to moisture deficit conditions. This steep CO_2 gradient enables high rates of photosynthesis to occur as temperatures rise. The second advantage of C4 photosynthesis is that a high concentration of CO_2 is maintained in the chloroplasts of bundle sheath cells, where it combines with RuBp and prevents photorespiration.

FIGURE 4.2 Differences between C3 and C4 photosynthetic pathways in the assimilation of CO_2 into carbon compounds

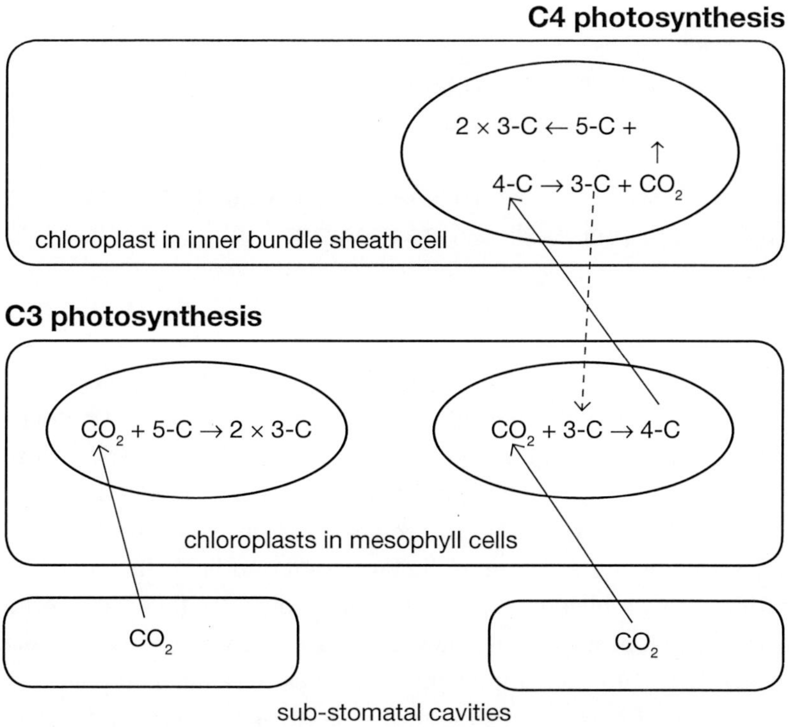

Translocation of sugars

Sugars are produced in the chloroplasts of leaf cells during photosynthesis. Tissues in different organs around the plant, such as roots and developing shoots, require sugar molecules for growth and energy. The plant is able to transport sugars such as sucrose from the leaves to other parts of the plant in phloem conducting tissue. The sucrose is loaded into phloem cells in the leaf and then translocated in concentrated phloem sap to tissues. This process requires energy, supplied in the form of ATP. It is useful to think of *source tissues* as those that export the photosynthate (the sugar product of photosynthesis) and *sink tissues* as those that use the photosynthate for metabolic processes.

Respiration

Plant cells throughout the plant require a continuous supply of biochemical energy for energy-consuming processes associated with growth and cell maintenance. The energy is supplied by the oxidation of 6-carbon sugars in a series of enzyme-controlled reactions known as respiration. This process is to some extent the reverse of photosynthesis, because it uses oxygen and produces CO_2 and energy. The yield of dry matter from a crop or pasture is the difference between carbohydrate gains from photosynthesis and losses from respiration. Respiration losses of carbon amount to roughly half the carbon fixed by photosynthesis, and this can increase to 70–80 per cent of daily carbon gains by photosynthesis in tropical regions due to high night temperatures, which increase night-time respiration rates.

The complete oxidation of a 6-carbon sugar involves three different biochemical processes in different parts of the plant cell. The first process is called glycolysis. In this process the sugar is converted into 3-carbon molecules of pyruvic acid in the cytoplasm of the cell. During glycolysis some of the energy in the sugar is released by the formation of ATP and NADH. The second process is a series of reactions that form the Krebs' cycle or tricarboxylic acid (TCA) cycle, in which pyruvic acid is oxidised to produce three molecules of CO_2. This takes place within small organelles in the cytoplasm called mitochondria. The released energy is also stored as additional ATP and NADH. In the final process oxygen is consumed and the hydrogen and extra electrons in NADH are transferred by cytochrome molecules to form water and more ATP. In addition to supplying ATP and reductant compounds in the form of NADH, respiration also supplies carbon-containing molecules that are intermediates in the breakdown of sugar molecules. These are able to act as 'carbon skeletons' for the synthesis of other important molecules in plant metabolism, such as amino acids and fatty acids. Respiration is controlled partly by environmental factors such as temperature, and in the case of plant roots, by the supply of gaseous oxygen. However, the respiration rate is also influenced by the plant's requirements for energy (ATP), the demand for reductant (NADH), and the need for carbon-containing molecules to synthesise new compounds in the plant cell.

Transpiration

The process of transpiration enables water to move from plant roots to the leaf canopy in the xylem conducting tissue. Water is essential for plant metabolism. It is the solvent for inorganic salts, sugars and organic anions in plant protoplasm and is the medium for all biochemical reactions. Water molecules adsorbed to the surfaces of ions and other solutes influence these reactions. The mass flow and diffusion of water through conducting tissues enables nutrient ions and metabolites to be transported upwards through the plant. The movement of water into cells results in the outward pressure on cell walls. This assists in cell elongation that follows the division of cells in meristematic tissue and provides rigidity and turgor for leaves, roots and other plant parts.

Transpiration is the loss of water vapour from plants. It is a passive process resulting from the evaporation of water molecules into the sub-stomatal cavities in leaves. There are small microscopic pores on the surfaces of leaves known as stomata. The rate of evaporation increases with temperature and the dryness of the air surrounding the leaves. Transpiration occurs through an uninterrupted water column that connects the water in the soil pores surrounding the roots, through the root epidermis, the root cortex, the xylem vessels of the root, stem, petiole, and leaf, and thence through the veins of the leaf to the moisture film in the walls of the mesophyll cells lining the sub-stomatal cavities in the leaves. When the stomata in the leaf surface open to enable CO_2 to diffuse inwards for photosynthetic assimilation, water vapour in the cavities is able to diffuse outwards into the atmosphere. The water vapour in the cavity is replenished by water that evaporates from the cell walls lining the cavity, creating a negative pressure or tension in the water in nearby leaf veins. Because of the powerful cohesive forces between water molecules themselves and the capillary forces attracting water molecules to the surfaces of the small conducting vessels, the water column that stretches back to the soil solution cannot be broken. So water flows up from the soil solution to replace the moisture that evaporates into the sub-stomatal cavity and diffuses out into the atmosphere. The evaporative process is valuable in that it helps to cool the leaves that are exposed to high ambient temperatures. The stomatal pores provide the main pathway for transpiration because the upper and lower surfaces have an impervious waxy layer called the cuticle. Stomata close as the plants experience a water deficit. They therefore help to protect the plant against excessive water loss, and assist in regulating the water balance of the whole plant. The supply of CO_2 for photosynthesis ceases when the stomata are closed. Plants therefore make a huge compromise where CO_2 assimilation proceeds at the expense of unavoidable moisture loss.

Nutrient uptake

Mineral nutrient ions are another environmental resource that plant roots must acquire from the soil solution. Mineral nutrients make up a small proportion of the fresh weight of plant tissue, with the balance consisting of water and organic material made up of carbon, oxygen and hydrogen. The mineral fraction consists of 12 mineral elements or nutrients that are essential for the plant to grow and complete its life cycle. Plant nutrients are macronutrients or micronutrients, depending on their concentrations in plant tissue. (Macronutrients include nitrogen, potassium, calcium, phosphorus, sulfur and magnesium; micronutrients include iron, manganese, boron, zinc, copper and molybdenum.) The concentrations in plant tissue adequate for plant growth vary considerably between species and depend on plant age, the plant part and environmental conditions.

There are two stages in the process of roots taking up nutrients from the soil solution. In the first stage, nutrient ions move to the root surface. These ions move principally by the processes of mass flow and diffusion. Mass flow involves the transport of nutrient ions that are dissolved in the soil water as it moves to the root surface during the transpiration process. It is responsible for the movement of ions that are abundant in the soil solution such as calcium (Ca^{2+}) and magnesium (Mg^{2+}) ions, and perhaps the nitrate (NO_3^-) ion in high-nitrogen soils. The phosphate ($H_2PO_4^-$ and HPO_4^{2-}) and potassium (K^+) ions are present at very low concentrations in the soil solution so mass flow cannot deliver them to the root surface in sufficient quantities to meet plant demand. Their concentrations in the soil solution therefore decline at the root surface. This creates a gradient of increasing ion concentration away from the root. Ions away from the root can diffuse towards the root surface down this concentration gradient by their random thermal motion in solution.

The second stage of nutrient uptake involves the movement of ions from the root surface or epidermis, through the root cortex. The movement of ions into root cells is an active process, as the ion concentration is substantially higher in the xylem sap compared with the soil solution. Root respiration is required to supply energy for ion uptake. This depends on an adequate supply of oxygen to the roots.

Roots of different plant species differ in their capacity to take up nutrient ions from the soil solution because they have different morphological, biochemical and biological adaptations. For example, grasses are able to develop extensive finely branched fibrous roots that explore large volumes of soil to assist in nutrient uptake. Root hairs can develop just behind root tips and these increase the surface area for the absorption of nutrient ions. Many plants also are able to become infected by mycorrhizal fungi, which produce hyphae that extend out into the soil matrix, to increase the surface area available for nutrient uptake. Other features include the excretion of organic molecules from the root tips, such as organic acids. These dissolve insoluble compounds in the soil and this releases ions for nutrient uptake. Other plants can release enzymes from their roots to catalyse the breakdown of organic molecules that release inorganic ions for plant uptake. An example is the release of phosphatase enzymes that help break down organic molecules containing phosphorus.

Control of growth processes by plant hormones

Plant hormones are substances that play a dominant role in regulating plant growth and development. They are produced in one part of the plant and are transported in the phloem or the xylem (or both) to *receptor sites* where they affect or regulate different plant processes. The three groups of hormones that tend to stimulate growth and development are the cytokinins, gibberellins and auxins, while a fourth hormone, abscisic acid, tends to restrict growth. For some plant hormone effects, additional secondary compounds are required at the sites of action. The modes of action are often complex, and it is not fully understood how these hormones operate or how environmental conditions influence their synthesis and the sensitivity of receptor sites.

The impact of roots on the growth of plant shoots is mediated by plant hormones. Root meristems produce cytokinins. The xylem translocates the cytokinins to the shoots, where they stimulate new growth. The expanding leaves of the shoot apex stimulate gibberellin synthesis. This encourages stem growth. This means that a large vigorous root system will result in increased shoot growth. Cytokinins are produced in the shoots as well. Further research is required to understand the role of cytokinins transported to the shoot from the root, relative to those synthesised in the shoot.

Abscisic acid (ABA) is a plant hormone that plays an intriguing role in controlling a plant's response to water deficit stress. As the soil dries, ABA is synthesised in the roots and transported in the xylem to the leaves. At the same time ABA synthesis in the leaves increases. Water stress also increases xylem pH, and this changes the ABA molecule in the leaf, increasing its movement into the guard cells of the stomata, causing them to close, and prevent excessive water loss from the plant. In addition to closing the stomata, ABA increases root growth relative to shoot growth, and assists the roots to transport water to the shoots. Both of these effects help the plants cope with dry soil.

Environmental determinants for plant growth

Temperature

Temperature extremes limit where particular plants grow, as each plant has an optimal temperature range for growth. Temperate crop plants such as wheat, barley or canola are tolerant of frost when they are in their vegetative form, so they grow during the winter in frost-prone areas. Other crops such as peanuts, sorghum or maize are killed by frosts. They are grown during frost-free times of the year. These latter plants are known as *summer* crops while the former group are known as *winter* crops. Similarly, frost-sensitive 'tropical' grasses and legumes grow in regions free of frosts. Sugar-cane requires a relatively warm (>15°C) frost-free environment and is therefore restricted to the tropical or sub-tropical coastal areas. Temperature also has a controlling influence on the initiation of flowering for some plants that grow in cold climates. These plants require a period of low temperatures to enable their growing points (shoot meristems) to change from producing vegetative tissue to producing reproductive tissue that will develop into flowering inflorescences. This effectively delays the time of flowering until the spring. The delicate flowering organs therefore are not damaged by frosts, but appear instead in the warmer spring months when successful pollination is possible. This induction of flowering by periods of low temperature (usually 2–8°C) is called *vernalisation.*

Light

Sunlight also influences plant growth in a number of ways. The first is that a higher light intensity increases the growth rate of the plant, particularly where the leaf canopies of crops and pastures are dense and there are shaded lower leaves. Even though the light intensity required for maximum photosynthetic rates in C3 plants is usually much less than that of full sunlight, the presence of lower shaded leaves means that the photosynthetic rate for their whole leaf canopy will continue to increase at light intensities above those that maximise photosynthesis in the upper leaves. These principles also apply to C4 plants except that very high light intensities are generally required to attain 'light saturation' because of the C4 plant's ability to maintain CO_2 concentrations in the chloroplast. The second way that light influences plant growth and development relates to the length of the day and how this affects flower initiation. For example, short winter days will delay flowering in some winter cereals until a critical minimum length of day is reached, when flowering is initiated. These plants are *long-day plants* and the response helps to synchronise their flowering with the spring. Like vernalisation it can be seen as an adaptation to cold seasons to ensure that plants do not flower during the cold winter. Some summer-growing crop plants like soybeans are *short-day plants*, in which the initiation of flowering only occurs after a maximum length of day has occurred. This ensures that the plants will flower at the appropriate time in late summer.

Soil water supply

Soil water supply has a major impact on the growth of crop and pasture plants in Australia because of the frequent occurrence of dry spells during the growing season. During these spells the plants experience a water deficit stress because transpiration demand exceeds the rate of supply of water from the soil, and this stress reduces plant growth rates. Moisture stress affects the root system and the effect then impinges on the shoots. It is not surprising, therefore, that crop and pasture yields are directly related to the amount of rainfall received during the growing season. Excess soil water accumulating in the root zone as a result of poor drainage reduces crop and pasture growth. This production constraint is quite widespread in southern Australia, where some soils drain poorly and winter rainfall exceeds evaporation rates. Plant roots in these conditions are unable to obtain oxygen for root respiration so root function declines and there is a concomitant decline in the rate of plant growth.

Mineral nutrient supply

Chemically infertile soils where plant roots are unable to take up sufficient quantities of mineral nutrients to meet the demands for growth pose major constraints on crop and pasture growth. In many Australian soils the concentrations of plant-available nitrogen and phosphorus in the soil solution are low, and an increase in their supply will increase plant growth rates. Plant response to increases in the supply of deficient nutrients follows the law of *diminishing returns*. As more and more of a limiting nutrient is supplied to the plant, the growth response to extra nutrients is less and less. Australian farmers can correct deficiencies in nutrient supply by using fertilisers. However, fertilisers are expensive inputs so increased plant production is required to make the expenditure worthwhile. Mineral nutrients are one of the few essential environmental resources for plant growth that can be manipulated by the farmer.

Soil fertility and plant growth

In this section, the properties that contribute to soil fertility are explained briefly. An understanding of the parameters of soil fertility is essential for understanding the agricultural production process. Several very comprehensive specialist texts about soil science in the Australian environment are listed at the end of the chapter.

Components of the soil matrix

Four major components make up the soil matrix. These are the mineral fraction of the soil, soil organic matter (including the living organisms in the soil), soil water, and soil air. The organic and mineral fractions are intimately associated and form a porous solid matrix; around half of this matrix consists of pore space. The pores are filled by water and air. The water is called the soil solution and contains dissolved inorganic and organic solutes, while the soil atmosphere consists primarily of nitrogen and oxygen, with a higher concentration of carbon dioxide than the atmosphere above the ground.

The mineral fraction of the soil

Soils are formed by the weathering of rocks on the earth's surface. The main products of rock weathering are the naturally occurring primary minerals that are resistant to weathering, such as particles of quartz; the smaller particles of secondary minerals such as clays and oxides of iron and aluminium; and metal ions that dissolve in the soil solution. Mineral particles are graded in different size fractions according to their diameters, as outlined in table 4.1. Soils are given a texture class according to the relative amounts of sand, silt and clay that they contain. For example, one scheme designates a sand as a soil that contains >85 per cent sand; a loam as a soil that contains 50–70 per cent sand, 20–40 per cent clay and 10–25 per cent silt; and a clay as a soil consisting of >40 per cent clay, <25 per cent silt and <60 per cent sand. There are 11 soil texture classes ranging from silty loams to sandy clays. The texture of a soil is a stable property of that soil and cannot be altered by management. A measure of texture known as 'field texture' can be easily measured in the field by wetting a small amount of soil in the palm of your hand and kneading it into a ball and then rolling it into a ribbon between your thumb and forefinger. A sand will not form any ribbon, a loam will form a ribbon around 25 mm long while a heavy clay will form a ribbon at least 75 mm long.

Texture is one of the most important properties that determines the way the soil can be used for agriculture. For example, fine and medium-textured soils (such as clays and clay loams) are able to store moisture and nutrients in much larger amounts than the coarse sandy soils, and are therefore more productive. By contrast, sandy soils allow water to pass into and drain through the soil, and so have benefits for irrigation. They also require less energy for cultivation than heavy clay soils, which are sticky when wet and hard when dry. In general, the most desirable soils for agriculture (except for flooded rice) are loams. Loams contain significant proportions of sand, silt and clay; they have good moisture storage capacities yet do not have the low permeability and difficult cultivation characteristics of clays.

TABLE 4.1 Mineral particles in the soil matrix

PARTICLE NAME	PARTICLE DIAMETER (mm)
Gravel	greater than 2
Coarse sand	2–0.2
Fine sand	0.2–0.02
Silt	0.02–0.002
Clay	less than 0.002

Soil organic matter

Soil organic matter consists of a mixture of carbon-containing compounds (other than carbonates) that are derived from plant litter, dead roots, animal and microbial residues and their excreta, together with new substances formed by soil organisms. The amount of organic matter in soils varies from less than 1 per cent by weight to greater than 95 per cent in the case of peat soils. Most surface soils in Australia contain between 2 and 7 per cent organic matter, with the amount decreasing to less than 1 per cent in the deeper subsoil. The amount of organic matter in a soil reflects the balance between carbon inputs from plant and animal residues and excreta and carbon losses from decomposition by soil organisms. Factors reducing the decomposing activities of soil organisms, such as low temperature, poor drainage and strongly acidic conditions, lead to an accumulation of organic matter in soils.

The decomposition of organic residues in soil results from the oxidising activities of soil organisms to produce their energy, water and waste CO_2. Part of the energy is used in the process by soil organisms for their maintenance and growth. In the process, mineral nutrients are released as inorganic ions from complex organic molecules, a process known as mineralisation. The released ions include nitrate, sulfate, ammonium, and phosphate ions. These are available for plant uptake, or they might be taken up by soil micro-organisms and be unavailable for plant growth. Nutrient uptake by micro-organisms is known as nutrient immobilisation. The process of decomposition also leads to the formation of a relatively resistant fraction in the soil organic matter known as *humus*. Humus is a dark, fine-textured material. This is a mixture of modified resistant compounds in plant residues, such as lignin and waxes, and new compounds produced by micro-organisms.

Organic matter is of immense importance to soil fertility. It makes the soil matrix a more favourable medium for root growth, in a number of ways. These include stabilising the structure of soil aggregates; acting as a reservoir of nitrogen, sulfur and phosphorus for plants; increasing the retention of nutrient cations on the negatively charged surfaces of organic colloids; increasing the moisture storage capacity; and providing substrates for biological activity in soils. Indeed the maintenance of organic matter concentrations in the soil can be seen as one of the key sustainability indicators for agricultural production. The subject of sustainability is discussed in chapter 10.

Soil organisms

Although the soil organic matter consists of the remains of plant and animal life, it would be wrong to view the soil as just a storehouse of dead cells. Rather, it contains a large number and variety of living organisms, such as small animals and micro-organisms that feed upon the organic materials in the soil. These living organisms constitute the soil biomass that makes up less than 5 per cent of the total soil organic matter content. The small animals include invertebrates such as earthworms, slugs, beetles, mites, termites, springtails, centipedes, ants and insect larvae. The earthworms, because of their size and activity, are perhaps the most important group consuming plant residues. Of the soil micro-organisms, including bacteria, fungi, actinomycetes, and protozoa, the bacteria are the smallest (1–4 μm in length) and most numerous. Soil fungi are less numerous but their biomass can be similar to that of bacteria because of their thread-like growth forms. The protozoa are the smallest of the soil animals (5–40 μm in length). These feed on bacteria and fungi, are universally present and their biomass is comparable with that of the earthworms. The earthworms are more mobile within the soil; most species burrow between the surface litter layer and the soil to ingest plant remains and mineral soil.

The numbers of soil organisms relate closely to the amount of organic matter in the soil. This means that the largest numbers are in the soil surface, and their numbers decrease rapidly with depth. Micro-organisms reside in soil pores but their size determines the pores they can inhabit. This means there are few organisms in the smaller micropores (<0.2 μm in diameter). These micropores are therefore sterile. Bacteria, actinomycetes and fungi inhabit the larger mesopores (0.2–30 μm in diameter), where they are generally free from the predation of larger protozoa. The largest soil pores of all, the macropores (>30 μm diameter), are habitable by most soil organisms.

Soil organisms make a major contribution to soil fertility. Bacteria and fungi that obtain food and energy from organic material in the soil are termed *heterotrophic*. Heterotrophic bacteria and fungi are the major decomposers in the biosphere. They are able to liberate mineral nutrient ions from the organic substrates and make them available again for plant uptake. Heterotrophic bacteria and fungi therefore have pivotal roles in the recycling of nutrients. This is an essential process for the survival of both natural and managed plant communities. Another type of bacteria, called autotrophic bacteria, derive their energy by oxidising inorganic material. These play key roles in increasing plant-available nitrogen and sulfur. For example, the nitrifying bacteria are responsible for oxidising ammonium nitrogen to nitrate nitrogen. This is the preferred form of N for most crop plants. Other bacteria can oxidise reduced forms of sulfur to plant-available sulfates. Further types of micro-organisms form invaluable symbiotic relationships with crop and pasture plants; for example, the *Rhizobium* bacteria are able to fix atmospheric N_2 in the root nodules of leguminous plants. Certain soil fungi form mycorrhizal associations in or on plant roots that improve phosphorus uptake by the plant root in low-phosphorus soils. They also assist in zinc uptake. Finally the earthworm has beneficial effects by incorporating organic material and surface-applied soil amendments such as lime or phosphate rock into the root zone, as well as creating burrows that improve water and air movement and root penetration through the soil.

Some soil organisms have detrimental effects on plant growth. For example, the cereal cyst nematode (*Heterodera avenae*) and the take-all fungus (*Gaeumannomyces graminis*) severely damage the roots of cereal crops and require specific rotation sequences to decrease their impact. The root lesion nematode (*Pratylenchus* species) and the common root rot (*Fusarium* species) damage wheat plants. *Rhizoctonia solani* is a root fungus that damages the roots of many field crops. Cane grubs, soldier fly and false wireworms damage sugar cane plants. The phytophera soil fungus (*Phytophera clandestina*) is responsible for root rots in subterranean clovers.

Soil water

The soil can be viewed as a reservoir of water for plants that grow in the soil. Rainfall is usually somewhat irregular in agricultural regions of Australia. This means that the capacity of a soil to absorb rainwater and store the water for subsequent use by plants during dry periods has a direct bearing on the productive capacity of that soil. The water is held within the soil matrix in the network of soil pores that occupy the spaces between soil particles or aggregates. The nature of the soil pore system depends on the soil texture, the soil aggregation, the organic matter content, and burrowing activities of soil macrofauna. These features determine how much water the soil can store for plant growth and how rapidly the soil can absorb rainfall.

Immediately following a period of extended rainfall or irrigation the soil surface is completely saturated. This means that all of the pores in the soil surface are filled with water. Subsequently the water in the macropores (diameter >30 μm) that cannot be held against the force of gravity, drains through these large pores. A stage is reached where only the smaller pores with diameters less than 30 μm will be holding water. The soil is then said to be at 'field capacity' in that it is holding the maximum amount of water that can be held against gravity; this would occur two days after heavy rain or irrigation, provided that there is no impediment to free drainage and there has been no water lost through surface evaporation. Following this, in the absence of further water inputs, there will then be a progressive loss of soil water over time from plant transpiration or soil evaporation. As each amount of water is lost, the remaining water is in progressively smaller pores in closer proximity to the finer soil particles. This water is held more 'tightly' by the soil because of increasing adhesive forces between water molecules and soil particles. Eventually the plants can no longer take up water so they will wilt and eventually die from lack of moisture if no further rain falls. The soil water content is then at the 'permanent wilting point': the mesopores (diameter 0.2 μm – 30 μm) have been emptied, and the remaining soil water is located in the micro-pores (<0.2 μm diameter) and is unavailable to plants.

Of interest here is the amount of water that plants can use as the soil dries. The total amount of water in soil is measured in the laboratory by drying the moist soil in an oven at 105°C and expressing the weight of water as a percentage of the weight of oven-dry soil. This measure is known as the gravimetric water content. Another measure is the volumetric water content. This is the volume of water expressed as a percentage of the volume of soil. These measures do not tell us how much of the soil water is available to plants; this is a function of the porosity and associated water-holding forces within the soil matrix. Soil texture is one soil property that has a major impact on the available water capacity of a soil (the amount of water available for plant growth that a soil can store, that is, the difference between field capacity and permanent wilting point). Sandy soils have a high proportion of macropores. These drain readily, resulting in a low available water capacity, whereas clays have many meso and micropores, resulting in high field capacities and wilting points. Loams generally have the highest available water capacities because of their larger proportion of mesopores.

An important concept concerning soil water is the tension of soil water in the soil matrix. This involves the concept of *soil water potential.* Soil water potential is a measure of the negative pressure plants must 'pull' against to draw the soil water into their roots. A number of factors contribute to the total tension or water potential that holds water in the soil. These include the matric potential. This results from the forces of adsorption between water and soil surfaces and capillary forces by which water is held in small pores; the osmotic potential, where solutes in the soil solution create an osmotic pressure that attracts water; and the gravitational potential, where gravity exerts a tension on the soil water that is drawn upwards into plant roots. The magnitude of the negative value of the sum of the matric, osmotic and gravitational potentials is called the *total water potential.* The more negative the soil water potential, the more strongly water is held in the soil and the more difficult it is for water to move from the soil to the plant roots and from the roots to the plant tissues.

One important property of the soil matrix relating to water is its rate of water infiltration. This is a measure of how easily irrigation water or rainwater enters the surface layers of the soil. The infiltration rate has major implications for agriculture, because if rain falls at a greater rate than it is able to infiltrate into soil, it runs off the land and can result in soil erosion. Water can move through larger pores more quickly than smaller pores: in fact the rate of flow of water through a saturated pore is proportional to half the pore diameter (i.e. radius) raised to the fourth power. Sandy soils with a high proportion of macropores because of the large size of sand particles, have higher infiltration rates than clay soils, which have a high proportion of very fine micropores. Other factors that reduce infiltration rates are the soil water content (dry soil takes in more water than wet soils) and the presence of a compacted, dense soil layer. Also the process of soil dispersion when the soil has a high exchangeable sodium content, results in the soil particles dispersing and blocking soil pores, and thereby reducing infiltration.

The notion of the soil being a reservoir of plant-available water leads on to the 'bucket concept'. Here, the larger the bucket the greater the amount of plant-available water that can be stored in the profile for use by crop or pasture plants during the growing season. Quantifying bucket size can help us identify more productive from less productive soils for dryland agriculture. This can be done by measuring the soil water in the total rooting zone in the profile, when the profile is full of soil water, perhaps with the use of extended drip watering. This measure is known as the upper drained limit (UDL). A second measure of profile water content is required after the crop has grown and taken up all the soil water available to it. A plastic rain shelter is needed to ensure that no rainfall is allowed to enter the profile late in the season. This dry measure is known as the crop lower limit (CLL). The difference between the UDL and the CLL represents the size of the bucket, and is the amount of water that the crop has been able to remove. The technical term is the plant-available water capacity or PAWC of the profile. The challenge facing Australian farmers is how they might increase PAWC to make better use of limited rainfall.

Soil atmosphere

Living organisms and plant roots within the soil matrix require a supply of gaseous O_2 for their aerobic respiration. The O_2 is normally contained in and provided by the soil air that exists in pore spaces not occupied by soil water. Air porosity is the volume of soil air expressed as a percentage of the soil volume. This is inversely related to soil water content: the air porosity increases as the water content of the soil decreases. Respiration occurring within the soil will proceed at a rate determined by temperature, water and substrate availability, so long as O_2 from the atmosphere can diffuse into the soil matrix to replenish the O_2 in the soil air at a rate fast enough to meet respiratory requirements. This process of gas exchange between the soil air and atmosphere is known as *soil aeration*.

The problem of reduced aeration occurs in a wet, poorly drained soil where most of the pore space is filled with water. Such soils generally have few pores large enough to drain freely when the soil is at field capacity. The rate of gas diffusion, and therefore the rate at which O_2 can be supplied to sites of active respiration, is 10,000 times higher in air compared with water. Soil respiration in these wet soils therefore leads to a depletion of O_2. This in turn limits the root activity of plants sensitive to O_2 deprivation. Low air-filled porosity and poor aeration are likely to occur in clay soils that contain few macropores or that are low-lying, in permanently saturated soils.

The soil air contains much higher concentrations of CO_2 than the atmosphere, which consists of 0.03 per cent CO_2. In the soil, CO_2 is normally ten times this figure (0.3 per cent), and in wet soils can reach concentrations in excess of 1 per cent. As O_2 becomes deficient, the number of aerobic bacteria decreases and are replaced by anaerobic bacteria. These are able to use substances other than gaseous O_2 as electron receptors for respiration. The soil matrix generally becomes less favourable for root growth due to the formation of toxins such as ethylene and soluble manganese ions, and to the loss of plant-available nitrate nitrogen. This is formed into gaseous N_2O and N_2 by the process of denitrification.

Organisation of the soil matrix

Soil colloids

Soil colloids are the smallest physical particles in the soil matrix, having diameters between 2 nm and 1000 nm (1000 nm = 1 μm). Their small size means that they have a large surface area relative to their mass or volume, and this makes them very reactive. Indeed, the reactivity of soil colloids enables them to have a major impact on the physical and chemical properties of the soil, and therefore on the fertility of the soil. The two major colloid groups contained in the soil matrix are (1) the inorganic colloids comprising the secondary clay minerals, and the hydrous oxides of iron and aluminium, and (2) the organic colloids produced by the humification process. In general the inorganic clay colloids tend to have more effect on the soil matrix than do the organic colloids.

The importance of colloids on soil chemistry results from the development of negative or positive charges across their surfaces, which attract cations or anions respectively from the soil solution to become adsorbed onto the surface. Without this adsorption the ions would be vulnerable to leaching loss from the soil by drainage water before they could be taken up by plant roots. Negative charges occur on the surface of clay minerals when cations of lower valency (that is, a lower positive charge) replace higher valency cations within the clay molecular structure. Negative charges also develop on the surfaces of clay minerals, hydrous oxides, and humus or organic colloids, when the soil becomes more alkaline as hydrogen atoms dissociate from exposed groups (i.e. the –OH group) on the molecular surface. A positive charge develops on the surface of some clay minerals and oxides in acid soils when hydrogen ions become attached to exposed groups.

The important effect that clay mineral colloids in particular have on the physical fertility of the soil relates to the proportions of exchangeable cations that are adsorbed to their surfaces. The negatively charged clay surface in moist soil is surrounded by a swarm of attracted cations, with the concentration of cations being highest at the surface and decreasing away from the surface. The cations thus form a diffuse electric double layer, referred to as the Gouy double layer after the man who described it. Now, the thickness of this layer decreases with the concentration of electrolyte (dissolved salts) in the soil solution, and with the valency of the adsorbed cations.

What might this mean in practice? Take the case of a sodic soil, where adsorbed univalent sodium (Na^+) ions exceed 6 per cent of the cation exchange capacity. When this soil becomes wet, the wide double layer associated with the univalent sodium ion and the low ionic concentration causes the clay particles to separate and form a stable colloidal suspension. The particles disperse because they cannot approach each other close enough for the forces of attraction (van der Waals forces) to overcome the repulsive effects of the wide, positively charged double layers. The dispersed clay will fill the pore spaces in the soil, reducing permeability to air and water, which can lead to waterlogging and poor aeration. In addition, water infiltration is restricted and a surface crust develops on drying.

FIGURE 4.3 (1) Cations and anions in the diffuse double layer at the surface of a negatively charged clay colloid, and (2) differences between coagulated colloids in a non-sodic soil and the dispersed colloids in a sodic soil

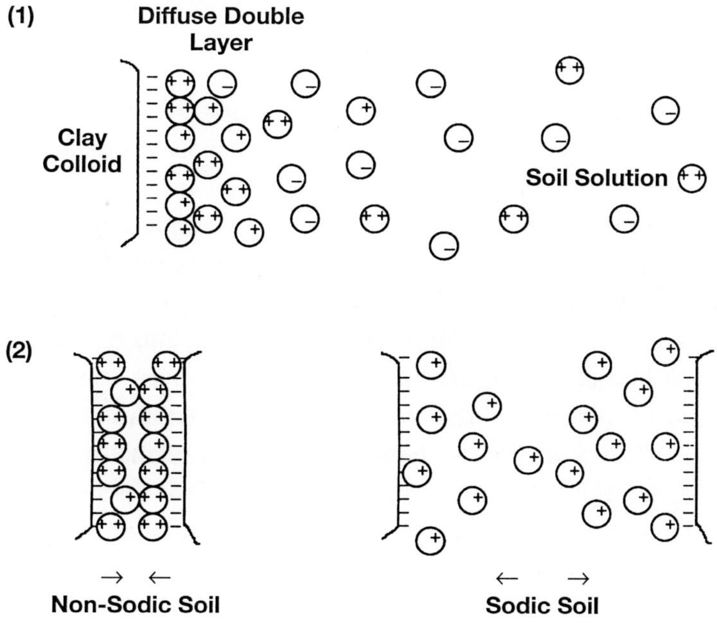

Soil structure

Soil structure refers to the arrangement of the major primary soil particles—the clay, silt, and sand fractions—into larger units of various sizes called aggregates, and the associated network of pore spaces that is subsequently formed between the aggregates. The resulting porous matrix is able to provide a favourable environment for plant roots to grow and penetrate, for storing water, and for supplying gaseous O_2 to roots and soil organisms. Maintaining a soil structure that is favourable for root growth is one of the important goals in soil fertility management.

Soil aggregation results from natural processes. Some processes break up or disrupt the soil mass into smaller particles, while others bring smaller particles into close contact to encourage aggregate formation. The disruptive forces result from wetting and drying cycles. These cycles are important for clay soils that respond by swelling and contracting. The process of freezing and thawing has similar effects. Soil water expands as it cools from 4° to 0°C. Plant growth also contributes as it assists in soil drying. The wetting and drying of clays leads to the formation of flat clay particle combinations known as domains. These are up to 100 μm long and 1–10 μm wide. Aggregate formation begins when the domains are drawn together in the first instance by flocculation, and with other colloidal material by the attraction of positive and negatively charged surfaces, and by van der Waals forces and hydrogen bonding, to form micro-aggregates (<250 μm diameter). Micro-aggregates are themselves bound together and bonded with sand and silt particles to form water-stable macro-aggregates (>250 μm diameter) by the binding effects of organic materials such as microbial polysaccharides, polyuronide gums, root-tip mucilage, mucus from soil fauna, and fungal hyphae.

It is not possible to measure soil structure directly, but it is possible to measure other closely related soil properties that give a good indication of how favourable the structure might be for root growth. The first is the porosity of the soil. This is the total pore space occupied by air and water. A fertile topsoil has a total porosity of around 50 per cent. The second is the soil's bulk density. This is the mass of a unit volume of dry soil. It gives an indication of the looseness or compactness of the soil and hence its porosity. Root penetration becomes more difficult as the soil matrix becomes denser, but the critical bulk density that reduces root penetration depends on soil moisture and texture; dry bulk densities above 1.5 g/cm^3 in clay soils and above 1.7 g/cm^3 in sandy soils are likely to reduce root penetration unless they remain moist. A third measure relating to soil structure is the percentage of water-stable aggregates, which is the percentage of aggregates that do not fall apart in water.

Soil profile

Soil layers, and the materials that formed them, lie above the bedrock at the surface of the earth's crust. A given soil may have developed from the weathering of that underlying bedrock, or it may have formed from material transported to that site by wind, water, gravity or glacial ice, or by the decomposition of vegetation. The soil layers usually differ in colour, but in some cases there are differences in texture, structure or hardness. The layers are termed *soil horizons* and the set of horizons at a given location is known collectively as the *soil profile*.

FIGURE 4.4 The profile of a red-brown earth soil in northern Victoria. A dark-red B horizon overlies a paler yellow-brown C horizon.

The uppermost horizon of a soil profile is often described as the topsoil and is called the A horizon. It is higher in organic matter than other layers in the soil profile, because organic matter accumulates from the deposition of plant residues on the surface. This gives rise to a darker colour than the soil in the lower horizons. The percolation of rainwater through this surface horizon removes soil materials in solution or suspension, by the process of eluviation and so this layer is described as *eluvial*. Materials that are removed by this process may include clay minerals, oxides of iron and aluminium, or salts such as calcium carbonate or calcium sulfate. The underlying subsoil, containing less organic matter, may accumulate materials washed down from the A horizon and this accumulation is termed *illuviation*. The illuvial layer in the profile is referred to as the B horizon. Lying beneath the B horizon is the relatively unaltered material that formed the soil layers. This is known as the parent material and is designated as the C horizon.

Soil profiles vary markedly across the landscape, and different profiles are rarely separated by sharp boundaries. Soils can also be dynamic bodies that change over time, as profiles develop. Thus there can be variation in the soil mass in both space and time. Similarly, the soil matrix on a smaller scale is not uniform and there can be variation in its properties over a distance ranging from millimetres to metres. This spatial soil heterogeneity or variability means that effective sampling protocols need to be used in collecting a soil sample that is representative of an area.

Soil fertility and infertility

Soil fertility is a measure of the capacity of the soil profile to support productive plant growth. A fertile soil implies that the matrix provides a favourable environment for plant roots to grow and take up water, mineral nutrients, and oxygen; all essential for the plant to grow. Visualising the characteristics of a fertile soil can provide a checklist against which other soils can be compared. In the following sections the focus is on those aspects of the matrix that make it a favourable or unfavourable environment for root growth and function.

Restrictions on root growth

The first requirement for a fertile soil is that root extension and proliferation are not restricted by factors within the topsoil or subsoil as long as the water and temperature regimes are favourable for growth. The plant growth hormone cytokinin is synthesised in the meristems of root tips: increased root growth means more root tips, more cytokinin exported to the shoots, and greater shoot growth. There are numerous restrictions on root growth in many agricultural soils in Australia. These include soluble aluminium and manganese ions that can occur in high concentrations in the soil solution of acidic soils. The aluminium ions restrict cell division in the root tips of sensitive species like barley, resulting in short stubby roots. There are also some toxic substances, such as high concentrations of sodium and chloride ions in saline soils; phytotoxic organic chemicals released from the decomposition of some crop residues; and gaseous ammonia can occur in alkaline soils when ammonium-releasing fertilisers are sown with or near the seed of a crop.

FIGURE 4.5 Extensive root proliferation through a well-structured soil matrix, where there is little restriction to root growth

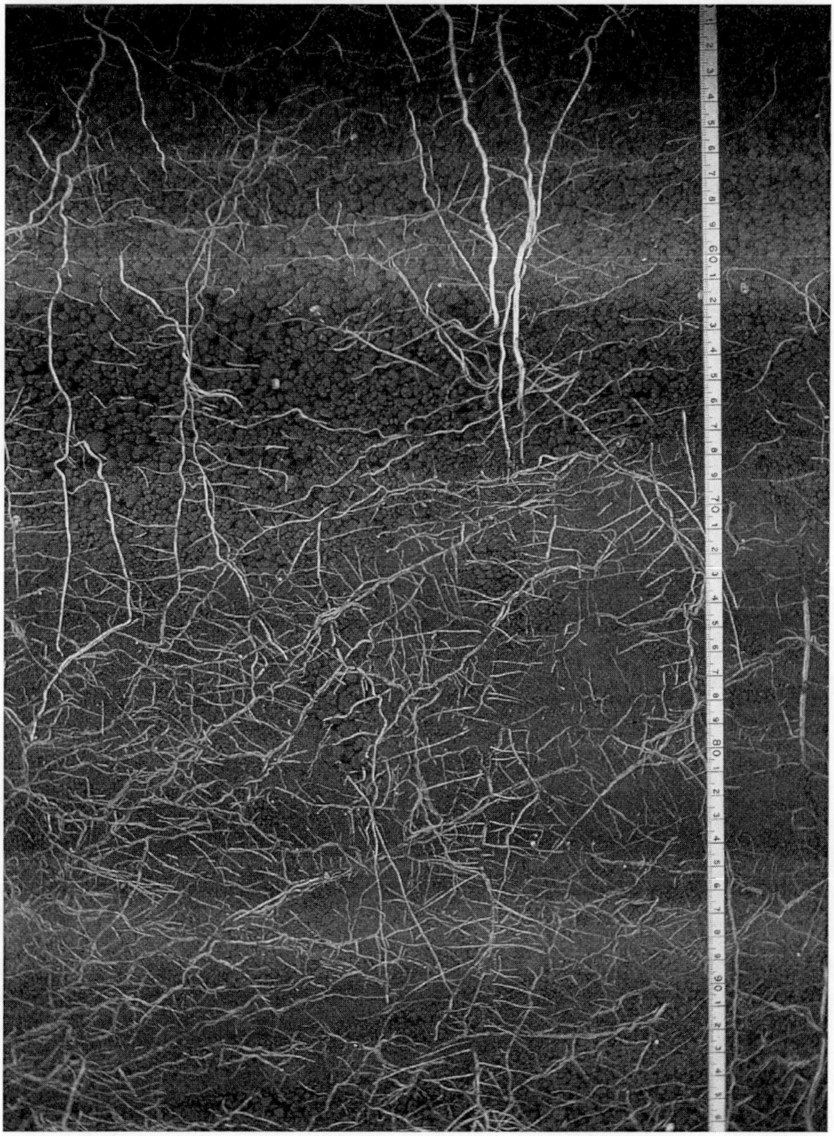

Root growth is also restricted in poorly aerated soils because of the lack of gaseous oxygen for root respiration. This can occur in soils with low porosity when they remain wet for extended periods. Similarly, restrictions occur in very dense soils due to mechanical resistance to root penetration. These soils with high bulk densities have few macropores, so roots have to displace soil particles in order to penetrate through the soil. This can require considerable force. Root growth decreases as the soil becomes drier and as soil strength or mechanical impedance increases. There are some soils with very unfavourable structure, such as the very dense B horizons of some sodic soils, where root growth is restricted by low O_2 supply under wet conditions and by mechanical impedance as the soils become drier. Root growth can only occur over a narrow range of water contents in these infertile soils. Infection of plant roots with root pathogens in the soil will also restrict root growth. Sensitive cereal crops that are sown in cropping paddocks with a history of infection by *Rhizoctonia*, cereal cyst nematode, or root lesion nematode will most likely become infected and suffer yield reductions due in part to stunted root growth.

An increased research focus during the first decade of the twenty-first century has increased awareness of another problem facing Australian agriculture. This is the extent of subsoil constraints that restrict root growth in the subsoils of many regions of the wheat belt. Restricted root growth means that subsoil water cannot be used by the crop or pasture plants. This limits the plant-available water capacity of the profile, and so reduces the productive capacity of the soil. There are a number of constraining agents in these subsoils including salinity (an excess of salts), sodicity (an excess of Na^+ ions), boron toxicity, acidity (excess of H^+ ions), and the physical constraints of high bulk density and low porosity. It now appears that in the drier areas of the wheat belt, the major constraint is subsoil salinity, which is virtually impossible to ameliorate. The challenge for farmers and scientists is to identify the constraint(s) and the areas where it/they are operating, and develop strategies that minimise the impact on farm profitability.

Mineral nutrient supply

A fertile soil must provide and maintain adequate concentrations of mineral nutrient ions in the soil solution for uptake by plant roots. The two considerations here are the actual concentration of the nutrient ion in the soil solution (the *intensity factor*) and the ability of the soil's solid phase to continue to replenish ions in the solution as they are removed by root uptake (the *quantity factor*). Both factors depend on the nutrient reserves in the soil horizons.

Nitrogen is the most important limiting nutrient for non-legumes in Australian soils. Soil nitrogen reserves are held in the soil's organic matter (particularly microbial and plant residues) that must be converted by soil micro-organisms into the plant-available ammonium (NH_4^+) and nitrate (NO_3^-) ions. The supply of these ions therefore depends on the soil organic matter levels, the nitrogen concentration within that fraction, and the existence of favourable conditions for mineralisation to proceed. Nitrogen (as nitrate) is also readily lost from the soil by leaching and denitrification under anaerobic conditions. This reduces the nitrogen supply for plants. Phosphorus is also a major limiting nutrient in Australian soils due to their old age and extensive weathering history. Few soils in Australia have primary minerals, such as apatite, that contain phosphorus. The soil phosphorus reserves are stored in the organic matter or as adsorbed or insoluble inorganic forms. There is a small pool of plant-available phosphorus in the soil solution that has been released by mineralisation or is in equilibrium with less strongly adsorbed phosphorus on soil particle surfaces.

Sulfur, like nitrogen, is stored in the soil organic matter and can be released as the sulfate form (SO_4^{2-} ion) for plant uptake. Some inorganic sulfate may be adsorbed onto positively charged sites on clay and oxide surfaces in acidic soils. Sulfur (as sulfate) often accumulates in deeper soil layers of acid tropical soils as a result of this process.

The three essential macronutrient cations, potassium (K^+), calcium (Ca^{2+}), and magnesium (Mg^{2+}), are stored in the soil as exchangeable cations. These are adsorbed on negatively charged sites on clay and organic colloid surfaces. Soils that have high cation exchange capacities and are highly saturated with these major cations will therefore be well supplied with them. Other reserves include primary silicate minerals, such as the feldspars, micas, and ferro-magnesian silicates, that have not been decomposed by weathering. These primary minerals will weather in time to release essential cation nutrients so soils rich in primary minerals such as some basaltic soils are well supplied with these basic cations.

The trace elements Fe, Zn, Mn, Cu, B and Mo, which are required by plants in small amounts, are also contained in a range of primary minerals that are released on weathering. These nutrients will also occur in a variety of forms in the soil, such as in the biomass, adsorbed onto exchange sites, and also in combination with insoluble or soluble organic molecules. Soil conditions influence their availability to plants. For example, molybdenum often becomes deficient as the soil pH_{CaCl_2} falls below 4.5–4.8. Zinc deficiency can develop in alkaline clay soils, while some of the calcareous alkaline soils in South Australia are deficient in manganese for sensitive crops. A number of organic soils are copper deficient because copper can bind with organic matter. The insoluble forms of these trace elements are generally retained in the soil profile, thus maintaining soil reserves. However, widespread deficiencies have developed in some of the old, weathered sandy soils in coastal parts of eastern and southern

Australia due to cumulative leaching losses over time. Copper deficiency has also developed in some acidic sandy soils in Queensland that are used for sugar production.

Water relations

A fertile soil acts as a large reservoir of plant-available water in both the topsoil and the deeper subsoil. This means that the soil matrix must contain a large proportion of intermediate-sized mesopores that will retain the soil water at water potentials between field capacity and permanent wilting point, and that this desirable porosity extends through a deep rooting zone from the surface horizons to the subsoil. There should also be no impediments to prevent roots from growing into the subsoil to access the deeper water reserves. Soils that have the highest mesoporosity are those with a loam-type textural class and a well-developed soil structure.

The second key requirement for favourable water relations in a fertile soil is the capacity for a high rate of infiltration into the soil. The presence of larger macropores will facilitate the entry of rain or irrigation water into the soil surface layers and then enable it to continue to move down through the rooting zone even when the soil becomes saturated. Another requirement for effective infiltration is the existence of a favourable soil structure in the topsoil where well-developed water-stable aggregates do not disintegrate under the actions of raindrops or a rapid wetting. Unstable aggregates in the surface layers of the soil that collapse upon wetting will drastically reduce the infiltration rate. This disintegration is called *slaking* and results in the formation of a surface crust that reduces crop or pasture emergence and facilitates erosion of the surface soil. The productive capacity of sandy soils with high infiltration rates and low available water capacities, compared with clay soils with the opposite properties, depends to some extent on the rainfall patterns. When rain falls over an extended period to bring the clay to field capacity, then the clay will support plant growth longer into a dry period. Should the rain fall in frequent light showers after a dry period, then the sand will take in water to a greater depth. This enables the vegetation to continue to grow, while the clay will remain dry with no plant growth.

Aeration

For a soil to have adequate aeration, macropores must occupy more than 20 per cent of the volume of the soil matrix. These pores do not retain water at field capacity and are therefore filled with air, assuming that the surplus water is free to drain out of the soil. The network of macropores therefore provides air channels through a wet soil to enable gaseous O_2 to diffuse into the soil matrix and CO_2 to diffuse out. Sandy soils where the soil matrix consists of larger particles, have adequate macropores between the sand grains. When the air-filled porosity of a wet silt, loam or clay soil falls below 10 per cent, then the rate of O_2 supply is reduced sufficiently to limit the growth of most crop or pasture plants. An improvement in aeration therefore requires an increase in macroporosity. Improving soil structure and the degree of soil aggregation by increasing organic matter levels and avoiding soil compaction will lead to more macropores in the soil. Of course there are soils that are low-lying that become partially submerged by high watertables in wet times. In these conditions the macropores in the soil matrix are not able to drain freely so they remain full of water. As the gas diffusion rate in water is much less than that in air, then oxygen supply in these soils will become restricted and plant growth rates will decrease. Installing underground or surface drainage channels to remove the surplus water will improve aeration under these conditions.

FIGURE 4.6 A wheat crop that is experiencing waterlogging in a soil where impeded drainage prevents aeration through the soil matrix

Soil reaction

The soil reaction, which refers to the extent that a soil is acidic, alkaline or neutral, has a marked impact on soil fertility. Soil reaction is expressed in terms of the pH of the soil solution where pH is defined as the negative log of the hydrogen ion concentration in that solution. The optimum soil pH (measured in a soil / calcium chloride extract) for most crop or pasture plants is 5.5 to 6.0 because of the effect of pH on the availability of essential mineral nutrients. With increasing soil acidity (as the soil pH_{CaCl_2} falls below 4.7) the availability of all macronutrients declines. This is due to the reduction in the mineralisation of organic N, S, and P due to reduced microbial activity, the decline in P availability due to the formation of insoluble iron and aluminium phosphates, and the lower supply of calcium, magnesium, and potassium because the saturation of negative exchange sites by these cations diminishes with increasing soil acidity. In addition the availability of molybdenum also declines as the molybdate anion becomes increasingly adsorbed on soil surfaces below pH_{CaCl_2} 5.0.

Perhaps the major constraint posed by increasing levels of acidity in soils is the increase in the availability of toxic concentrations of aluminium and manganese ions in the soil solution. However, crop and pasture plants differ in their degree of tolerance of soil acidity. Forage legumes such as lucerne and annual medics are very sensitive to these toxicities, whereas *Lotus* species and the perennial grass cocksfoot are quite tolerant. Similarly, barley and canola are sensitive field crops whereas triticale and lupins are tolerant. As the soil becomes more alkaline and the pH increases above 8.0, there are some adverse effects on nutrient availability. For example, calcium ions react with soluble phosphate ions to form insoluble calcium phosphates. Similarly the solubility of iron, manganese, zinc and copper compounds decline and deficiencies of these micronutrients become more prevalent.

FIGURE 4.7 A plot of oats growing to the left of a plot of barley in an acid soil. Note the increased growth of oats on this soil compared to the very poor growth of the barley, which is very sensitive to the acid soil constraints

REVIEW QUESTIONS

1 What are the characteristics of a fertile soil?
2 What features of a soil limit plant growth?
3 How do C3 and C4 plants differ?
4 What are the main implications of different environmental characteristics for the management of different crops and pastures.
5 What processes are involved in photosynthesis?

FURTHER READING

CSIRO, *Soils: An Australian Viewpoint*, CSIRO, Melbourne, 1993.

Leeper, G. W. and Uren, N. C., *Soil Science: An Introduction* (5th edn), Melbourne University Press, Melbourne, 1993.

Loomis, R. S. and Connor, D. J., *Crop Ecology: Productivity and Management in Agricultural Systems*, Cambridge University Press, London, 1992.

Northcote, K. H., *A Factual Key for the Recognition of Australian Soils,* Rellim Technical Publications, Coffs Harbour, NSW, 1992.

Parsons, W. T. and Cuthbertson, E. G., *Noxious Weeds of Australia*, Inkata Press, Melbourne, 1992.

Pratley, J. (ed.), *Principles of Field Crop Production* (3rd edn), Oxford University Press, Melbourne, 1994.

White, R. E., *Introduction to the Principles and Practice of Soil Science* (2nd edn), Blackwell Scientific Publications, London, 1987.

WEBSITES

Australasian Farm Business Management Network: www.csu.edu.au/faculty/science/saws/afbmnetwork/

CSIRO: www.csiro.au/

National Farmers Federation: www.nff.org.au/

NSW Department of Primary Industry: www.dpi.nsw.gov.au/

Vic Department of Primary Industry: www.dpi.vic.gov.au/dpi/index.htm

5 Practice of Plant Production

CONTENTS

Introduction

The aim of the farmer is to grow a productive crop or pasture that will generate an adequate profit. Sound crop and pasture management is so important here in achieving the high yields that contribute so significantly to profit. There are certain management procedures required to achieve this outcome. The farmer has to choose the correct crop or pasture genotype to suit the climate and the soil conditions in the paddock. A vigorous and appropriately dense plant population must be established in the paddock. The plants must be protected from the competitive effects of weeds or the damaging effects of insect pests and diseases, and the physical and chemical fertility of the soil has to be made as favourable as possible for productive plant growth.

The key environmental factors that determine where a crop or pasture genotype can be grown are the rainfall (amount and distribution), the temperature regime, and in some instances the soil properties such as soil pH and its drainage status.

Agricultural field crops

Winter cereals

The four major winter cereal crops—wheat, barley, oats, and triticale—are annual grass crops that are grown during the cooler months of the year. Wheat and to a lesser extent barley are the real 'engine rooms' of the Australian grains industry. They are the number one and two crops respectively in terms of area, tonnage and value, and reward farmers when seasons are favourable and good management practices are employed. These crops are mainly sown in the May–June period and grow vegetatively through the winter. Floral initiation, under Australian conditions, occurs soon after seedling emergence in early winter, with heading (inflorescence appearance) and anthesis occurring later in the spring. Grain-filling proceeds during the late spring – early summer period when temperatures are increasing and soil moisture supplies are frequently declining.

The winter cereals and other winter crops face a number of climatic constraints in Australia. First, the hot and generally dry grain-fill period reduces the duration of the period, resulting in fewer, smaller grains and lower yields than would be achieved under cooler and more humid grain-filling conditions. Crops should not therefore be sown late or they might encounter these high spring temperatures and dry conditions during the grain-filling period. On the other hand, the crop should not be sown too early because the crop might experience a frost at the time of flowering. Frost drastically reduces grain yields. The farmer must therefore sow the crop at the optimum time. This generally occurs in the May to early June period. However, adequate rain must fall just prior to this optimum time to provide adequate soil moisture to germinate the crop seeds sown in the surface soil. This does not always happen. A delay in sowing will mean a later flowering, and the possible exposure of the crop to hot, dry conditions during grain-filling.

The winter cereals are grown in a broad area extending from central and southern Queensland, down through central NSW, northern Victoria, the southern portion of SA, and in a large area in the south-western quarter of WA. The growing area is limited towards the centre of the continent by low rainfall and in the higher rainfall coastal regions by unsuitable (sloping) topography, high rainfall, and more competitive land-use preferences. However, there has been an expansion of cereal cropping in the high-rainfall regions of the south-east and south-west of Australia, as farmers learn to grow crops during wet winters when water-logging limits crop yields. In the southern wheatbelt the rainfall has a winter–spring distribution. This is well suited to the needs of the cereal crop. However, in northern NSW and southern Queensland the rainfall is summer dominant, so successful cereal cropping in this northern wheatbelt relies on storing a portion of the summer rain in the soil to provide moisture for crop growth in the drier winter–spring period.

Wheat

Wheat (*Triticum aestivum*) is the major grain crop in Australia. It is grown in all states in the area known as the wheatbelt. More than 13 million hectares of wheat were sown in 2003 producing a crop in excess of 26 million tonnes, whereas less than 8 million hectares were sown in the early 1990s when subsidies from other wheat-producing countries depressed world prices. Drought years also reduce production, with only 10 million tonnes being produced in 2002 and 2006.

The average yield is low by world standards due in part to the hot, dry conditions as the crop matures. The lowest average yield of 0.77 tonnes per hectare occurred in the drought year of 1982/83 while the highest average yield of 2.11 tonnes per hectare occurred in 2001/02 when excellent seasonal conditions prevailed. Average wheat yields have been increasing over time with better agronomic practices such as improved rotations, reduced tillage, and sowing the crop on time. Plant breeders have also incorporated dwarfing genes that led to higher yields. The wheat crops grown in the northern wheatbelt (northern NSW and Queensland) experience milder winters and more rapid crop development, whereas winters in the south are generally wetter and colder with slower development. The clearer skies in the north increase the risk of late spring frost, and there is also more likelihood of wet weather at harvest in the northern regions. The harvests start first in Queensland in October and spread southwards to finish in Victoria in January.

The major attributes of wheat grain are the high concentrations of starch (60–70 per cent) and gluten protein (10–15 per cent). This makes wheat flour particularly suitable for bread making. The gluten proteins provide elasticity to the dough, and this enables it to rise with the entrapped fermentation gases, and a 'light' leavened bread results. Australian wheat is generally attractive to overseas buyers because of its whiteness (the dry harvest period permits white wheat cultivars to be grown), its ease of milling and low moisture content (average 9–11 per cent), and its cleanliness and freedom from insect pests. There is concern, however, about the lower protein concentrations in the grain that result from the declining nitrogen status of cropping soils.

Barley

Barley (*Hordeum vulgare*) is the second most important winter cereal in Australia after wheat. The production of barley doubled from 1992 to 2002 when more than 4 million hectares were sown and a record crop of 10.4 million tonnes was produced in 2003. Although it is grown in most areas where wheat is grown, its average yield of close to 2 tonnes per hectare is somewhat higher than that for wheat. The gross value of the barley crop is around $1 billion but this depends on the size of the crop and world prices, as the majority of the crop is exported.

Barley is either sold at a premium price for malt production or as feed barley for animal consumption. The premium can vary by $20 to $40 per tonne so farmers are really keen for their barley crop to be sold as malt grade. The malting process involves soaking the grain in water to make it germinate (producing diastase and maltase enzymes) and then drying the germinated grain in a kiln. The grain for malting needs to consist of large grains with a high germination (around 90 per cent) and a high starch, low protein content. This will result if the crop is sown early in favourable areas where the grain-filling period has mild temperatures and humid conditions. High rates of nitrogen fertiliser can cause unacceptably high protein concentrations in malting barley grain and should be avoided. Fortunately the proportion of the crop that is sold for malting has been increasing, with almost 40 per cent of the national crop now reaching malting grade.

The barley crop matures before other winter cereals so it is better suited to short-season low-rainfall areas. It also tends to be grown on lighter, more alkaline soils. Its earlier maturity also enables it to be grown in frost-prone areas on the farm, and in seasons when the crop is sown later due to delayed sowing rains.

Oats

Oats (*Avena sativa*) is a winter-growing cereal crop that produces a greater bulk of shoot growth than other cereals. It is therefore grown for grain and/or a winter fodder crop for sheep or cattle. The total area sown to oats varies from around 600,000 to 1 million hectares with average yields of 1.5–2 tonnes per hectare. Fodder crops are best sown in early autumn and provide autumn–winter feed for grazing stock. They can then be allowed to re-grow to produce hay or grain, or in a dry year can be grazed again. The majority of the oats grain is used domestically for stock feed. The grain retains a fibrous husk around the seed so it can be fed to ruminants and horses without any risk of digestive upsets that might occur with low-fibre grain. Production of oat grain is restricted to the southern wheatbelt because the plant is more sensitive to high spring temperatures than other cereals and because of the risk of crown rust outbreaks during the final stages of growth.

The oaten hay industry has developed strongly over recent years. This has been driven by the demand for quality hay, particularly by the livestock industries in Japan. The trade has grown from about 100,000 tonnes in 1993 to over 600,000 tonnes in 2005 and is worth about $150 million annually. Growing oaten hay for export requires good management as the price received is dependent on the quality of the hay.

Triticale

Triticale (*Triticosecale* species) is a synthetic hybrid between wheat and cereal rye that was first produced commercially in the late 1970s. Triticale as an alternative winter cereal has grown steadily, with the area fluctuating between 300,000 and 400,000 hectares each year. The grain is used mainly as an animal feed, as it has a similar feed value to that of wheat grain. The crop has found a niche as a winter cereal in environments where other cereals perform poorly. These include areas where acidic soils contain high concentrations of toxic aluminium, very cold areas or those with water-logged soils, and in regions with infertile sandy soils that are deficient in trace elements. New varieties with high yielding capacity, disease resistance and dual-purpose capability for grazing and grain, will increase the attractiveness of triticale as a dual-purpose crop. Although triticale flowers earlier than wheat crops planted at similar times, it takes longer to fill its grain, and so yield and grain size may be more affected than wheat by hot and dry finishes to the growing season.

Winter grain legumes

There has been a marked increase in the area sown to winter grain legume crops over the last 25 years. In the 1980s there was a seven-fold increase in the total grain legume production across Australia and this continued to increase during the 1990s. In the late 1990s more than 2 million hectares of winter grain legumes were grown across the Australian wheatbelt. The main crops were narrow-leaf lupin, field peas and chickpeas, with significant areas of faba beans, lentils, vetch and white lupins also being grown. The main drivers for these increases in grain legumes were three-fold. The first was the need to grow 'break crops' in crop rotations between cereal crops to avoid soil- and stubble-borne diseases of cereals. The second benefit arises from the nitrogen fixation that occurs in the legume roots. This has been estimated to average around 160 kg N per hectare per year. A further driver was the need for crop diversification away from traditional cereals because of declining export prices associated with difficulties in international trade, and away from traditional cereal–pasture (livestock) cropping rotations in some areas.

Grain legumes are also grown and marketed as a profitable crop in their own right. They supply export and domestic markets where there is a growing demand for protein feed supplements for intensive pig and poultry industries. However, their relatively low grain yields and prices generally result in lower **gross margins** than cereals. They are also considered to be less resilient in dry years when lower legume yields are common. Choosing a grain legume crop to grow depends very much on climatic and soil factors.

Lupins

The widespread use of the narrow-leaf lupin (*Lupinus angustifolius*) as a grain crop became possible during the 1970s with the release of new, non-shattering, low alkaloid cultivars with suitable maturity times. More than a million hectares were sown to lupins from 1994 to 2000, producing in excess of a million tonnes of grain with an average yield of around 1.1 tonnes per hectare. The largest portion (around 80 per cent) of the national crop is grown on light-textured soils in WA.

The narrow-leaf lupin grain contains 30 to 35 per cent protein with high energy levels. This makes it a valuable feed supplement for sheep and cattle, and is also used in pig and poultry rations. It is relatively cheap to grow but is priced below other pulse crops due to a lower concentration of sulfur amino acids. As the domestic stock feed market is limited in WA, the majority of the lupins grown there are exported, and so the price is generally lower than $150/tonne. This price plus the relatively low yields of 1 to 1½ tones per hectare mean that the gross margins in WA have generally been less than $75 per hectare. The lupin price is higher in the eastern states, in excess of $250 per tonne, as the grain can all be sold locally as stock supplements. The gross margins in the east are still moderate and do not exceed $250 per hectare.

FIGURE 5.1a A crop of narrow-leaf lupins in flower in September

FIGURE 5.1b Lupin pods during the late podfilling period in late October

The crop is most suitable for districts where spring temperatures are mild and moist. High temperatures (>25°C) at flowering will result in flower and seed loss, and dry conditions in spring will reduce yields. Lupins also grow better where the soils are lighter in texture and moderately acidic. They should be sown earlier than cereals during the April–May period as frosting at flowering time is less of a problem. Late sown crops (June–July) run the risk of poor seed-set due to high temperatures at flowering.

Field peas

Field peas (*Pisum sativum*) are a significant winter grain crop being grown on around 300,000 hectares in SA, Victoria and WA. The majority of the crop is exported for human consumption in Asia and the Middle East and for stock feed in Europe. The pea crops generally have low yields from 1 to 2 tonnes per hectare; the price ranges from $200 to $230 per tonne; and the variable production costs of $200 to $300 mean that the crop gross margins will be moderate at less than $250 to $350 per hectare.

Field peas are generally sown at the same time as cereals during the May–June period, as the crop needs to flower after the last frost and fill their seed before high temperatures develop in the late spring – early summer period. The crop is more suited to heavier-textured soils with a neutral to alkaline pH but cannot tolerate poorly drained conditions. Field peas are seen as an alternative grain legume to lupins in cereal rotations. The advantages of field peas are that they can be sown later than lupins and they can be grown in a wider range of soils (particularly heavier soils). They also make greater winter growth, and are more suitable for hay. They are better suited to the higher rainfall areas of the southern wheatbelt.

Chickpea

Chickpea (*Cicer arietinum*) is a winter grain legume that is adapted to dry environments and requires higher spring temperatures to set seed. The crop is also well suited to alkaline, clay soils and so the crop is seen as a substitute for field peas on alkaline soils that are unsuitable for lupins. However, the chickpea is very sensitive to waterlogging and the root diseases that are exacerbated by poor soil drainage. From a humble beginning as a trial crop in Goondiwindi in 1978, there has been a marked expansion in chickpea production with the area sown increasing from over 6000 hectares in 1984 to more than 200,000 hectares in the early 1990s. However, outbreaks of Aschophyta blight in 1998 resulted in devastating crop losses. Breeding for disease resistance, initially with off-shore screening of lines in Turkey, led to the release of new resistant cultivars. The chickpea industry is now recovering with around 100,000 hectares being sown in 2004 and 2005, producing average yields of around 1.1 tonnes per hectare. The main production occurs in NSW, with smaller areas in Queensland and Victoria. The major portion of the Australian crop is exported, principally to the subcontinent of India.

Oilseeds

There is a strong demand for Australian oilseed crops. The consumption of vegetable oils for human consumption is increasing worldwide, and now there is an increasing demand for biodiesel—a transport fuel that can be produced from vegetable oil. In addition to this increasing demand for oil products that can be extracted from oilseeds, the demand for oilseed meal, the product remaining after the oil has been extracted, is increasing due to its effective use as a high-protein feed supplement. Local production of soybeans and sunflowers is not sufficient to meet the national demand, with significant quantities of soybean meal and sunflower oil being imported. On the other hand, Australia exports the majority of its canola grain and its cotton seed as whole grain.

Canola (Brassica napus, Brassica campestris)

Canola is now well established as the third most important winter crop in dryland farming systems in Australia. It is a winter-growing brassica crop that was developed by plant breeders in the 1960s and 1970s; they were able to reduce the levels of erucic acid and glucosinolates in seed of oilseed rape and produce a plant with seed containing 40–45 per cent of high-quality oil, consisting predominantly of unsaturated fatty acids. Extraction of this oil leaves a 30–36 per cent high-protein residue meal that is used for animal rations.

The rapid increase in the area sown to canola in the 1990s can be attributed to a number of factors. The first was the release of herbicide-tolerant varieties in 1993 that were tolerant to the herbicide triazine. This enabled farmers to control problematic broadleaf weeds such as wild radish (*Raphinus raphistrum*) and wild mustard (*Sinapis arvensis*) that compete with canola and suppress yield and contaminate the canola seed sample at harvest. The second factor was an increase in genetic-based resistance to the blackleg fungal disease (*Leptosphaeria maculans*), which had previously wiped out canola crops in Western Australia in the late 1970s. The final factor was the beneficial role of canola as an alternative profitable crop and its beneficial role in crop rotations. It acts as an excellent break crop for cereal root diseases such as take-all (*Gaeumannomyces graminis*) and cereal cyst nematode (*Heterodera avena*), with frequent reports of higher yields of wheat crops following canola, rather than following wheat. As a result of these perceived benefits, Australian farmers increased canola plantings to a peak of 1.85 million hectares in 1999 and produced a 2.2 million tonne crop. Since then the production has fallen to 1.4 million tonnes in 2005, with lower tonnages in the dry years of 2003, 2006 and 2007.

Canola is generally grown in more reliable, higher rainfall areas of the southern wheatbelt where spring temperatures are milder, as the crop is sensitive to hot, dry conditions at flowering. It is less sensitive to frost damage at flowering because the plant has an extended 20–40 day flowering period: a severe frost would only damage flowers present, rather than the whole inflorescence. Canola can therefore be sown earlier in the autumn than wheat, so that it flowers earlier in the spring, avoids hot, dry conditions during the pod-filling period, and can be harvested before the cereal crops. However, the uneven ripening of pods on the plant means that the crop has to be cut and windrowed, particularly in southern regions, to enable all pods to ripen without pod shatter and seed loss.

There are problems associated with growing canola. The crop yields less than cereals and yields are more variable. It requires more nitrogen and sulfur fertiliser and generally requires a greater investment in weed control. Risks of lower yields, due to dry finishes to the growing season, are greater. There are also risks of lower yields from infections with fungal diseases such as blackleg and sclerotinia (*Sclerotinia maculans*). Breeding has provided some genetic resistance to the former disease but has achieved less success with genetic resistance for sclerotinia.

A new era for canola is developing with innovations in breeding. The first involves the availability of genetically-modified (GM) canola varieties that are tolerant to non-residual, broad-spectrum herbicides. A number of these GM varieties were licensed in Australia in 2003 but could not be commercialised because of legislation enacted by all canola-growing states, prohibiting their use. Fears of negative biological impacts of GM crops and reduced marketing opportunities for the grains industry were the basis of these moratoria on GM canola. During the next four years support developed in the grains industry to remove the ban as these fears diminished. It is now likely that a rapid adoption of GM forms of canola will occur in Australia, as it did in Canada, where around 85 per cent of canola was GM in 2006. The great advantage that the new GM varieties offer will be the opportunity to sow the crops early and still control in-crop weeds with post-emergent, broad-spectrum herbicides.

The second genetic innovation for the canola industry is the availability of a canola-type crop that is adapted to low-rainfall areas receiving less than 350 mm of annual rainfall. Plant breeders have bred lines of *Brassica junceae* with similar oil and residual meal characteristics to canola, but with a greater tolerance to heat and drought. The crop is called Juncea Canola, and will provide farmers in the drier cropping regions of the Australian cereal belt with an additional crop choice.

Sunflower

Sunflower (*Helianthus annuas*) is a summer-growing, broadleaf plant from the family Asteraceae. It produces seed with a high oil concentration (40–50 per cent), in a circular flowering head at the end of a single stem. The high-quality, edible oil is rich in mono-unsaturated (oleic) and polyunsaturated (linoleic) fatty acids and is widely used in margarine manufacture. The high protein meal (35 per cent protein) remaining after oil extraction is used as a livestock feed supplement. Around 150,000 to 200,000 hectares of sunflowers, with a value of around $50 million are grown each year to meet the local oilseed demand. However, the dry years since 2000 have reduced the area sown to sunflowers to less than 100,000 hectares per year.

The crop is widely grown in Australia in areas where there is sufficient summer rainfall or irrigation. The major sunflower areas are therefore in central and southern Queensland and northern NSW, while smaller areas are grown under irrigation in southern states. Sunflowers are tolerant to frost in both the seedling and seed-ripening stage, which means that they can be grown in the early and late part of the summer. Although the plant is relatively tolerant of moisture stress during the vegetative stage the seed yields are sensitive to moisture stress at flowering and early in seed development. High temperatures during seed development will also reduce oil quality. Relatively moderate gross margins of around $300 per hectare can be expected with dryland sunflowers for yields around the national average of 0.9 to 1.0 tonnes per hectare. The high oil concentration increases the price of sunflower to $575–625 per tonne.

FIGURE 5.2 A late crop of sunflowers at the late flowering stage in early March

Soybeans

Soybeans (*Glycine max*) are a summer-growing legume crop that produces seed with 20 per cent oil, and over 40 per cent high-quality protein. The crop is grown for its oil content but the high-protein meal, remaining after the oil is extracted, is a valuable protein supplement that is used in the pig and poultry industries. Soybeans require high temperatures and an adequate moisture supply to produce high yields. In addition, the plant is very sensitive to the daylength. This varies with latitude, and the temperature regime. Soybean cultivars must therefore be adapted to the particular region where they are grown. In order to produce maximum yields, the soybean plant needs to develop an adequate leaf canopy before flowering and then have sufficient time, while temperatures remain high, to complete pod-filling. The crops generally yield from 1½ to 3 tonnes per hectare with on-farm prices of $350–400 per tonne, although some beans are sold at higher prices for human consumption.

The temperature and moisture requirements mean that they are able to be grown in irrigation areas in north-western and southern NSW and southern Queensland. Dryland crops occur in coastal river valleys and adjacent tableland areas of northern NSW, and in southern Queensland where there is adequate summer rainfall. The issue is that soybeans sold for oilseed only produce moderate returns and cannot compete profit-wise with cotton or rice. The area sown to soybeans is therefore restricted to between 50,000 and 100,000 hectares per year.

Summer coarse grains

Sorghum

Sorghum (*Sorghum bicolor*) is the most widely grown summer crop in Australia and ranks as the third biggest grain crop in terms of production. The grain is used mainly as a stockfeed. The major portion of the crop used to be exported to Japan, but the expanding feedlot industry in Australia is now consuming much of the national crop. A shortfall in production is looming because of the increasing demand by animal industries for feed grain. The national crop since 2000 has been around 2 million tonnes, except for the dry years in 2002–03 and 2006–07, when smaller tonnages were produced.

Because the crop is tolerant of moisture stress and high temperatures (due to its control over leaf stomata and to its extensive root system) it is grown in the hotter drier parts of the northern wheatbelt in northern NSW and southern and central Queensland where rainfall is adequate. The average yield is around 2–2.5 tonnes per hectare and this reflects the relatively harsh production environment, when moisture stress during grain-filling is a common constraint. It is often grown in these areas as an opportunity crop when the lack of rain has prevented wheat sowing. It also provides a useful rotation crop to control weeds such as wild oats in wheat.

FIGURE 5.3 A dryland crop of sorghum during late heading stage in north-western NSW in late March. Note the low plant density and the abundance of summer weeds.

Maize

Maize (*Zea mays*) is a summer-growing crop that is sensitive to high temperatures and moisture stress, particularly during flowering. It is therefore confined to the wetter areas of the coast and tableland regions of Queensland and NSW and inland irrigated areas of these states as well as southern NSW and northern Victoria. The profitability of the crop is yield dependent. Average dryland crops can yield 3–4 tonnes per hectare with gross margins of $300–$400 per hectare, depending on price. Irrigated crops can average around 10 tonnes per hectare with a gross margin of around $1000 per hectare. However, top irrigated yields on farm have exceeded 20 tonnes per hectare with good management.

Production of maize fluctuates between 300,000 and 400,000 tonnes of grain. It is linked to the domestic market with only a small surplus being exported. About 70,000 tonnes of grain are produced locally for human consumption, while the balance is used as stock feed. In recent years there has been an increasing interest in growing maize for silage production for both dairying and the beef feed-lot industries. This trend is likely to continue.

Rice

Rice (*Oryza sativa*) is a summer-growing cereal, requiring relatively high temperatures (25–30°C) and adequate water to produce high yields. It is grown under flood irrigation in the Riverina region of southern NSW. The long, hot sunny days during the summer months enable well-adapted medium- and long-grained rice varieties to produce high grain yields. At around 10 tonnes per hectare the yields of Australian rice are the highest in the world. The crop can be profitable with gross margins in excess of $1000 per hectare. However, the area sown to rice is tightly regulated and the land must have deep impervious clay subsoils to prevent drainage losses.

The national rice crop expanded to around 1.3 million tonnes in the late 1990s but then contracted sharply in the 2002 drought, and there was practically no rice grown in 2007 due to the continuing dry weather and the lack of irrigation water. This highlights how the rice industry can be dramatically affected by climatic events. Rain run-off is required for irrigation dams so that water can be allocated for rice growing. This means growing rice can become an opportunistic activity, carried out intermittently when there is water available. A second issue relates to temperature. Hot weather is required during ear formation; if cold southern air is forced into southern NSW at this time, then yield losses can occur. This happened in 2005 when the average rice yield fell by around 40 per cent to only 6.6 tonnes per hectare.

FIGURE 5.4 Harvesting rice in April in the Murrumbidgee Irrigation Area in southern NSW

Cotton

Cotton (*Gossypium hirsutum*) is a high-value and profitable summer crop with gross margins for average irrigated crops often exceeding $1400 per hectare, when average prices are obtained. Australian cotton farmers are highly skilled and have learnt how to produce the highest yields in the world. In 2005 the industry yield was a world record at 9.2 bales per hectare. This was three times the average world yield. Not surprisingly the cotton crop has experienced a marked expansion since the early 1960s, with more than 500,000 hectares being grown in the 2000–01 season, contributing well in excess of $1 billion worth of lint and seed. The unfortunate run of dry seasons from 2000 to 2007 caused a substantial decline in cotton production, with only 140,000 hectares being grown in 2006.

Cotton is grown for the fibres that are attached to the fruiting bodies (bolls). The harvested fibres are taken to a cotton gin, which removes the seed from the fibres. The seeds are then extracted for their oil content. The crop is sensitive to moisture stress, particularly during the bud and boll formation phases; it also requires warm temperatures over the growing season and cannot be sown until soil temperatures in the spring exceed 16°C. Cotton is therefore grown principally under irrigation from central Queensland south to the Murrumbidgee valley, with major production occurring in the Namoi, Macquarie, and Gwydir valleys in NSW and on the Darling Downs and in the

Emerald district of Queensland. In recent years there has been an increase in the production of dryland cotton in summer rainfall regions.

A real issue for cotton growing is the plant's attractiveness to insect pests, particularly to native and cotton budworm larvae (*Helicoverpa* sp.). Controlling these pests required the use of repeated applications of insecticide. Alarm bells were ringing in the mid 1990s with reports of resistance developing to insecticides, and to insecticide contamination occurring in rivers and livestock. Changes came quickly with the development of genetically modified varieties that incorporated the Bt gene that expresses a protein in the cotton leaves that is lethal for the budworm larvae. Use of these GM plants combined with integrated pest-management practices have reduced insecticide use to about 5 per cent of its use in the early 1990s.

FIGURE 5.5 Harvesting cotton in the Trangie district of central NSW in April

Horticultural cropping

Commercial vegetable and fruit production involves significant capital investment in land and machinery and high labour costs. Fruit and vegetable crops are usually perishable, and pests, diseases and bad weather can rapidly reduce potential returns from the crop. Harvesting usually has to be completed quickly. Vegetable growers usually have a choice of different ways to sell their product. It might be sold direct to supermarkets, or to wholesalers or to processors or to a grower cooperative or directly to customers in central markets. Prices of vegetables fluctuate because of their perishability, seasonal patterns, climatic conditions causing fluctuations in supply, and demand being fairly fixed. Horticultural tree crops have a limited life and need replanting periodically. There is also usually a time lag between time of planting and time of bearing the first product. These crops have an initial capital investment in establishment and irrigation system, and annual costs to nurture and protect the plants and to harvest the crop.

Growers of fresh vegetables share the following production, harvesting and marketing decision problems with broad-area crop producers:

- which crops and combination of crops to grow
- which variety of crop to grow
- which husbandry practice for each crop
- which rotation to use
- when to plant
- how best to fully utilise permanent labour and how much and what quality of casual labour is needed
- how best to ensure timeliness of operations, such as sowing, weeding, disease control, spraying and harvesting
- how much to invest in specialised machinery.

There are also special problems of importance to vegetable growers:

- how to obtain the best technology to maximise profits in a highly intensive, risky and disease-prone production system
- where and how to sell the produce
- what to do with the produce if prices are low
- what the likely price will be at harvest
- whether to plant all the area of crop at the one time or to stagger plantings
- when to have produce ready for market to obtain the best result.

Vegetable growers face some major uncertainties that make planning difficult:

- the price of produce sold varies often daily and from corresponding periods of one year to the next
- yields vary seasonally, often because of management, as well as pests, disease and water supply
- planting and harvest times vary considerably if there is unsuitable weather and there are implications for other crops that are planned
- it is difficult to know how much to invest in preventing adverse effects from rare but serious weather events such as frost
- the amount of labour required, and when, varies with the size of the harvest.

As in broad-acre crop production, the success of horticultural cropping is determined to a great extent by the quality of the management, especially timing of the major husbandry operations.

Pasture plants

The native grasses of Australia developed under unique environmental conditions. The soils were low in fertility and the grazing pressure from macropods such as kangaroos was generally light. Also, the native grasslands were subject to repeated burning by Indigenous Australians. The early European settlers noted that in the absence of regular burning these grasses were relatively unproductive and matured rapidly to produce seed. This resulted in a rapid decline in their nutritive value for sheep or cattle. There were then rather extended periods when the grasses provided dry feed of low quality, before fresh growth commenced in the following growing season. Not surprisingly, the level of animal production on these native species was quite low, with improved performance occurring when the cyclical pattern of fresh growth occurred.

The development of pasture improvement practices from the 1930s to the 1970s brought about substantial increases in animal production in the higher-rainfall regions of the country. The process involved improving soil fertility with single superphosphate and replacing the native (and naturalised) species in grazing paddocks with productive legume and grass species. These are more productive under the higher stocking rates and improved soil fertility regimes that occur with pasture improvement. During the last fifty years breeding and selection programs have improved the performance of the introduced pasture species, and there are now a large number of introduced pasture species available to the Australian farmer. In this section the focus will be on the attributes of these plants, which are important when making choices for specific pasture environments. Most emphasis will be placed on species for temperate regions, where pasture improvement has been a particularly successful farm practice.

Tropical species

There are two groups of tropical pasture species for growing in the tropics (latitude <23.5°S). The first group includes species that are grown in the wet tropics where the rainfall exceeds 1500 mm and there is no pronounced dry season. The search for suitable perennial tropical legumes for this area has been disappointing. Those that have been introduced have a trailing habit with long sprawling stems that can climb up companion grasses. However, these legumes have been shown not to persist under the high stocking rates that might be theoretically possible, given the high rainfall figures for the region. Perennial grasses used for intensive grazing in the wet tropics include species like signal grass (*Brachiaria decumbens*). Fertiliser nitrogen is used on signal grass pastures in the absence of a suitable legume.

The second group of tropical species comprises those that are grown in the seasonally dry tropics where the rainfall is usually less than 1000–1200 mm and there is a pronounced dry season. Here the approach is to oversow native pasture with a range of tropical legumes such as *Stylosanthes* species. These species do not require fertiliser at planting to ensure establishment, so oversowing can be a low-cost option. However, on low phosphorus soils, the application of phosphorus fertiliser or the provision of phosphorus supplements during the growing season can improve animal performance. In wetter areas closer to the coast, legumes like Glen joint vetch (*Aschynomone americana*) are showing promise, although some fertiliser might be required for optimum production. Away from the coast the tree legume leucaena (*Leucaena leucocephala*) is being used on well-drained, non-acidic soils. This shrub produces high-quality leaf material that is valuable during the dry season when the quality of the native pastures decline.

Subtropical species

The subtropical species are grown in more inland regions of central Queensland and parts of northern NSW, above latitude 30°S. This group includes mainly perennial grasses that are drought resistant and capable of withstanding frosts. Buffel grass (*Cenchrus ciliaris*) is a drought-tolerant species that has a number of cultivars suited to the drier inland regions with rainfalls between 350 and 900 mm. Other grasses include Rhodes grass (*Chloris gayana*). A major deficiency with subtropical pastures is the lack of a suitable, grazing-tolerant legume. Temperate drought-tolerant legumes such as lucerne (*Medicago sativa*) are sometimes included, while winter-growing annual medics are sometimes sown on cropping lands in the region.

Temperate species

Temperate species are grown in temperate areas below latitude 30°S and are capable of withstanding frosts during the winter. Introduced species are considered in areas where there are relatively long growing seasons (greater than five months), because the lack of rainfall in drier areas would limit productivity.

The type of pasture used in the temperate regions of Australia depends very much on the 'dryness' of the summer period, with generally more productive, perennial species being grown where the summers are more humid.

One exception is lucerne (*Medicago sativa*), a drought-resistant, summer-active, perennial legume that can be grown in areas receiving as little as 250 mm a year, and also in high rainfall areas where the rainfall exceeds 1000 mm. This valuable forage legume has a deep tap-root, which confers drought resistance. However, it can only be grown on neutral to alkaline soils that are well-drained and its erect habit makes it vulnerable to continuous grazing. Grazed lucerne stands will only persist if they are rotationally grazed to enable the plant to produce new leaves and replenish root reserves between grazings. Lucerne is also grown under irrigation for hay production where yields of up to 15 tonnes per hectare can be produced with repeated cuts in favourable areas. Interest in growing lucerne has increased appreciably in recent years because of the capacity of its deep roots to take up soil water from deep in the profile. This means that much of the rain that enters the soil will be prevented from draining below the root zone; it will instead be intercepted and taken up by the lucerne roots. Thus the plant can be used to manage dryland salinity.

Mediterranean or annual legume and annual grass pastures (very dry summers)

Pastures for the lower-rainfall areas of southern Australia, where the growing season lasts between five and eight months, are based mainly on annual legumes and grasses. These species are able to survive the long, dry summer as seed, produced each year in the late spring. A proportion of the seed produced is 'hard', which means that it will germinate over a number of seasons, thus ensuring the survival of the legume over time. Legumes include the annual medics such as barrel medic (*Medicago truncatula*) or strand medic (*Medicago littoralis*). These are grown on neutral or alkaline soils. They are widely used as legumes in pasture phases in rotation with crops in the drier, alkaline soil areas of the southern wheatbelt.

A very important annual legume in southern Australia is subterranean clover (*Trifolium subterranean*), which is adapted to moderately acidic soils, although one of the subspecies is adapted to neutral to alkaline soils. There are a range of cultivars for this legume, varying in the time required for flowering and seed setting, so subterranean clover can also be grown in the higher-rainfall areas with longer growing seasons. It is very well suited to heavy grazing pressure because of its prostrate, stoloniferous habit, and because its seed is buried beneath the soil surface (hence its name) as a result of the seed stalk turning down into the soil as the seed matures.

Annual ryegrass (*Lolium rigidum*) was frequently sown in this region prior to the 1970s. It germinates with the legumes after the opening autumn rains and sets large amounts of seed in the late spring. It has been widely used in pasture phases in cropping rotations and is widespread across the southern wheatbelt. However, it is a major weed in field crops because it can regenerate each year from seed produced the previous year. Furthermore, annual ryegrass has the propensity to rapidly develop resistance to selective herbicides because it has large genetic diversity and is out-crossing, and so requires pollen from another plant to set seed.

Annual legume/perennial grass pastures (dry summers)

In more humid regions (450–600 mm rainfall) where the growing season is longer (eight to ten months) it is possible to grow drought-resistant perennial grasses with subterranean clover. The grasses are phalaris (*Phalaris aquatica*) and cocksfoot (*Dactylis glomerata*). These species have deep root systems and are able to persist in a dormant state through the short summer drought. These grasses confer increased stability to the pasture sward. Soil conditions determine which grass is appropriate for a particular paddock: phalaris is best adapted to heavy soils and is tolerant of wet soil and flooding as well as heavy grazing pressure while cocksfoot requires a well-drained soil and persists better on sandy soils. Cocksfoot has a lower nutritive value than phalaris but copes better with acidic soils.

As the rainfall increases above 600 mm it becomes possible to grow perennial ryegrass (*Lolium perenne*) with subterranean clover. This grass is less drought resistant than phalaris or cocksfoot but is productive in autumn and spring and has a high nutritive value. Another perennial grass that is grown where there is more summer rainfall or on heavy soils where the summers are drier is tall fescue (*Festuca arundinaceae*). It is particularly well suited to wet soils and moderate salting. Its main drawback is the lack of winter growth. Breeding and selection is developing grasses that perform better in different grazing environments. For example, new cultivars of phalaris have been developed that can grow better in acid soils, while fescue varieties have been developed that are more winter active.

Perennial legume/perennial grass pastures (humid summers)

In still higher rainfall regions (>800 mm) where the growing season exceeds ten months, it becomes possible to grow white clover (*Trifolium repens*), a perennial legume. This very productive legume is suited to a range of soil types and fertility levels, and is capable of withstanding heavy grazing pressure because of its prostrate, stoloniferous growing habit. However, it has shallow roots that grow from stolon nodes, and this makes it sensitive to dry conditions and thus it is limited to areas with humid summers. Interest in white clover in high-rainfall areas of southern Australia has diminished somewhat in recent years as there has been a move towards the increased use of fertiliser nitrogen to boost grass growth in the pasture.

FIGURE 5.6 A sward of white clover and perennial ryegrass in the Tumut district of southern NSW in November

Crop or pasture establishment

Crops and pastures are managed plant communities that are grown in a particular paddock or area of the farm. A crucial phase in their culture occurs when the plant population is established in that paddock. This involves sowing the seed into moist soil after existing vegetation has been removed, to enable the seed to germinate, and emerging seedlings to establish in the soil. A major aim during this establishment phase is to provide conditions that favour rapid, vigorous growth of the establishing seedlings. Management is very important at this time because incorrect practices at sowing can result in a sparse stand.

Some crops like potatoes and sugar cane are established by vegetative propagation, while crops like tobacco have a very small seed and are established by transplanting the seedlings. However, most crop and pasture plants are established from seed so the following principles need to be followed:

1. The viability of the seed should be assessed and the seeding rate (seeds sown per hectare) adjusted accordingly.
2. The seed should be placed at the correct depth (depth is inversely related to seed size), as small-seeded species have limited seed reserves and can only emerge from shallow depths. For example, canola should be sown at 1–2 cm depth, whereas the larger soybean seed can emerge successfully from 3–5 cm depths.

3 Effective contact should be made between the moist soil and the sown seed, to enable soil water to move into the seed so that the seed can germinate. This requires a satisfactory fine soil tilth if the seedbed has been cultivated. Alternatively, the seed should be placed into a slot in moist soil with direct drilling, and the slot should be covered with loose soil to prevent moisture loss.
4 The correct seeding rate should be calculated to ensure that the desired plant density is established in the paddock. This will vary according to seed size, plant size and the likely growing conditions. For example, dryland sunflowers should be sown at 4–6 kg of seed per hectare to produce a population of 30,000–40,000 plants per hectare, whereas under irrigation the seeding rate per hectare should increase to 8–14 kg of seed to achieve a population of 60,000–100,000 plants per hectare.
5 The establishing crop or pasture must not be exposed to significant weed competition. Many studies have shown how yield reductions from weed competition are greatest when young seedlings have to compete with weeds for light, water or nutrients. For example, the small canola seedling is poorly equipped to compete with weeds, whereas the fully developed plant competes strongly by shading them.
6 Pests must be avoided as the young seedling is particularly sensitive to pest damage. For example, red-legged earthmites are most damaging to canola, lucerne and clover when the seedlings are emerging.
7 Nutrient requirements need to be considered to maximise early growth rates. Placing phosphorus fertiliser in a band close to legume seed will improve seedling growth in low-phosphorus soils because legume seedlings have poorly developed roots that restrict phosphorus uptake.

Protection from weeds, insect pests and diseases

Crops and pastures that have been selected for yield or nutritive value are attractive food sources for vertebrate and invertebrate pests. Frequent opportunities for weed invasion also occur when bare spaces develop within the managed populations. Crops and pastures must therefore be protected from the inevitable damage from pests and diseases if high levels of productivity and profitability are to be achieved. Failure to provide protection can result in costly outcomes for the farmer, including reduced yields from a reduction in plant numbers or plant vigour, a lower quality of the plant or animal product, increased production costs and, in some cases, reduced land values.

During the 1950s and 1960s there was a heavy reliance placed on the use of chemical pesticides to protect crops and pastures from the damage caused by weeds and pest animals (particularly insects). There was an optimistic belief that broad-spectrum, persistent biocides could eradicate these pests and provide clean, unblemished produce for the consumer. This approach inevitably led to secondary pest outbreaks, resistance to the more selective insecticides and herbicides, damaging effects on the environment and the occurrence of pesticide residues in the produce.

The era of chemical pesticides has now moved on to the era of 'integrated pest management' (IPM) (see chapter 11). The main tenets of IPM are (1) that the pest status (insect) in the crop or pasture is closely monitored, and (2) that the farmer does not intervene to control the pest until its population size reaches an economic threshold. Beyond this level, the potential injury level to the crop or pasture will warrant action to reduce its impact. The action that is taken involves a mixture of control measures, with a reduced reliance on the repeated use of selective chemical pesticides. There is a place for chemical pesticides but they are being augmented by cultural, biological, and genetic control methods. The use of chemicals will tend to be more strategic, with a preference for non-persistent types that have minimal impact on the environment.

The development of weeds that are resistant to herbicides has led to the adoption of integrated weed management (IWM) approaches. Here the aim is to avoid the repeated use of selective herbicides to avoid selecting for weeds that carry resistance genes for the herbicide. A whole suite of alternative weed control practices can be employed to destroy the weed or prevent it from setting seed in the paddock. These include green manuring, hay making, extra tillage, the growing of alternative crops at different times of the year, and the use of different herbicides to manage the weed. These strategies are often not as profitable as growing the crop, and so farmers begin to realise that there is a price to pay if weed populations build up and develop resistance to herbicides.

There are many examples of the different cultural, biological and genetic control strategies that are being used to protect crops and pastures. Cultural control uses regular farm management practices to reduce the activity of the targeted pest or disease by modifying the environment to favour crop or pasture growth, and at the same time discouraging the development of the pest population or the spread of the disease. An example is the use of break crops in rotations to reduce the incidence of the take-all root fungus in cereal crops. Biological control involves the manipulation of introduced or indigenous natural enemies of the pest in order to suppress it. One particularly successful example of biological control was the introduction from Argentina in 1925 of cactoblastis moths (*Cactoblastis cactorum*), whose caterpillar stage was able to control prickly pear species (*Opuntia inermis* and *Opuntia stricta*), which had severely infected some 25 million hectares of cropping and pastoral land in eastern Australia. Genetic control by breeding for resistance is particularly important for controlling plant diseases. An example is the continuing breeding programs that release new cultivars of wheat that are resistant to stripe and stem rust.

Managing soil fertility

Most Australian soils are not particularly productive in their native state. The soils have been developed from weathered and leached parent materials that were exposed on ancient land surfaces or were transported and deposited as erosion products from such landscapes. Consequently, many of the soils are deficient in some of the mineral nutrients that are essential for plant growth. Many are also low in organic matter due to the relatively low productivity of native vegetation. The surfaces of some soils are often hard-setting, resulting in crust formation on wetting. This may adversely affect plant establishment and water infiltration and increase the likelihood of erosion occurring when the vegetation is removed.

High-yielding, productive crops and pastures cannot grow on infertile soils. Enlightened soil management is therefore required to eliminate constraints to plant growth that result from soil infertility: the supply of limiting nutrients requires inputs of fertilisers; excessively acidic soils may need liming, to reduce the acidity; and soils with poor physical structure require special treatment to improve aeration and soil water retention, and to lessen severe mechanical impedance to root growth.

The principles of sound crop and pasture husbandry are well established. To grow profitable crops and pastures the farm manager has to combine and manipulate the plants, the natural environment and the economic and human factors. The economic return to managing crops and pastures well is very high compared to the cost of managing the farm resources poorly. Breaking the 'rules' of sound crop and pasture husbandry and management is the road to ruin.

REVIEW QUESTIONS

1 Explain the environmental determinants of plant growth.
2 What features of the plant's growing environment can be manipulated by the farm manager, and how?
3 What factors need to be considered when deciding on a sequence of crops to be grown in a particular area of land over a period of time?
4 Explain the roles that different types of crops and pastures can play in maintaining the soil in a healthy, productive state.

FURTHER READING

Lazenby, A. and Matheson, E., *Australian Field Crops: Wheat and other Temperate Cereals* (2nd edn), vol. 1, Angus and Robertson, London, 1987.

Leeper, G. W. and Uren, N. C., *Soil Science: An Introduction* (5th edn), Melbourne University Press, Melbourne, 1993.

Loomis, R. S. and Connor, D. J., *Crop Ecology: Productivity and Management in Agricultural Systems,* Cambridge University Press, London, 1992.

Lovett, J. V. and Lazenby, A. (eds), *Australian Field Crops; Tropical Cereals, Grain Legumes and Other Crops*, vol. 2, Angus and Robertson, London, 1987.

Parsons, W. T. and Cuthbertson, E. G., *Noxious Weeds of Australia*, Inkata Press, Melbourne, 1992.

Pratley, J. (ed.), *Principles of Field Crop Production* (3rd edn), Oxford University Press, Melbourne, 1994.

Wheeler, T. L., Pearson, G. T. and Robards, G. E. (eds), *Temperate Pastures: Their Production, Use and Management,* Australian Wool Corporation/CSIRO, Australia, 1987.

White, R. E., *Introduction to the Principles and Practice of Soil Science* (2nd edn), Blackwell Scientific Publications, London, 1987.

WEBSITES

Australasian Farm Business Management Network: www.csu.edu.au/faculty/science/saws/afbmnetwork/

CSIRO: www.csiro.au/

National Farmers Federation: www.nff.org.au/

NSW Department of Primary Industry: www.dpi.nsw.gov.au/

Vic Department of Primary Industry: www.dpi.vic.gov.au/dpi/index.htm

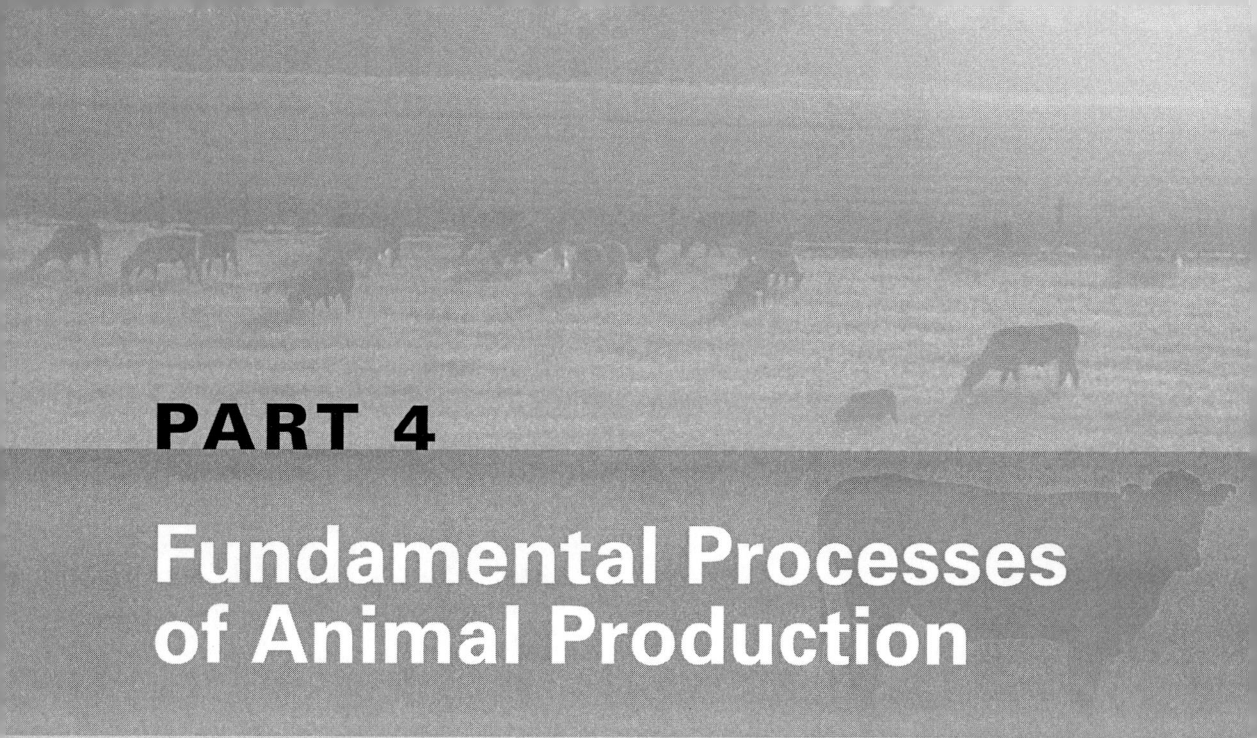

PART 4

Fundamental Processes of Animal Production

Principles of Animal Production

6

CONTENTS

Introduction

Animal production in agriculture in Australia involves extensive and intensive animal production systems based on grazing ruminant animals, and intensive animal production systems with non-ruminant animals housed and fed strictly formulated diets.

The purpose of a farm animal activity is to convert animal feed into saleable animals and animal products. This is done as part of a farm system, often involving other animal and plant-based farm activities, and is done to make a profit. The key factors in the operation and performance of an animal production system are keeping animals healthy, maintaining animal numbers, supplying them with the quantity and quality of feed they need at different times of the year, and improving the quantity and quality of saleable product they produce from the inputs used, for example feed, land and labour. In doing this, animal welfare is a continuing major consideration. The food and fibre produced has to be healthy, free of contaminants and meet the specifications for quality that are set by buyers in the market for the product. The main components and characteristics of animal production systems are shown in figure 6.1.

FIGURE 6.1 Animal farm systems

Inputs

Genotype
Nutrition
Disease
Physical environment
Animal management

Output factors

Amount of product, price
Value of product
Quality of product
Timeliness of supply

Components and characteristics

Animal species, breed
- Mature body size, body composition at age and weight
- Wool: mean fibre diameter and fibre diameter profile

Annual cyclic phenomena to be matched
- Feed supply and quality
- Plant growing season, pasture species, grazing management; supplementary feeding

Animal herd/flock cycles
- Age groups
- Fertility, seasonality of oestrus
- Fecundity, ovulation rate, seasonality of ovulation rate
- Gestation length
- Lactational capability
- Time to weaning, rate of growth to weaning
- Weight at weaning, growth rate post weaning

Land resource limitations
- Soil fertility
  - Plant water relationships, growth rate, mineral nutrient content
- Slope, aspect, drainage, etc.

Risk factors
- Variability in cycles
  - E.g. late season break, low rainfall, drought
- Disease
  - E.g. bacterial diseases, immunisation, footrot, fly strike, parasites; age and susceptibility; purchased animals, quarantine
- Market factors
  - Increasing costs, fluctuating value of animals and animal products

Productivity and product quality
- Meeting market criteria, e.g. carcass weight, fat cover; wool fibre diameter, tensile strength, dirt and organic fault
- Product premiums, e.g. continuity of supply, out-of-season supply

Animal functions

Farm animals in farm systems convert a raw material, animal feed, into a form that is valued either for producing more animals or is used directly for making end products that consumers want to buy such as meat, milk, fibres, and eggs. How does an animal function? An animal has a structure of bones, skin, muscle, fat and body fluids that are arranged into the following operating systems:

- a nervous system for sending instructions to all the parts of the body and for the animal to behave in a particular way
- an endocrine (hormone) system that regulates metabolic processes in different organs and tissues, and communicates with the brain and central nervous system
- a respiratory system that delivers oxygen to the blood stream and eliminates carbon dioxide to the atmosphere via the lungs
- a circulatory system for delivering oxygen and nutrients and hormones around the body and collecting unwanted products, such as carbon dioxide and wastes, and delivering the waste to the kidneys to be filtered out for eventual discharge from the body. As well, the blood helps to control body temperature by regulated expansion of blood vessels near the skin to enable heat to be dissipated by radiation or contraction to prevent heat loss
- an immune system operating at tissue and circulatory levels; for example, the blood contains white cells whose function is to fight organisms that cause disease
- a reproductive system comprised of specialised organs with processes integrated by hormones and operating through both physiological and behavioural mechanisms
- a digestive system that converts food components, that is, carbohydrates, fats, proteins, vitamins and minerals into simple substances that are in a soluble nutrient form and easily absorbed.

These systems underlie the orchestrated pattern of growth, reproduction and lactation. They are the internal systems through which amounts and qualities of product are determined and can be altered. Animal management is ultimately aimed at keeping these functions operating in an optimum range, and avoiding or controlling external factors that adversely disrupt them.

Digestive systems

A major distinction is made between animals that must have access to feedstuffs that are also directly useful to humans, and animals that can produce at high levels using solely plant materials that are of little direct use to humans or other animals. The latter class, herbivores, are adapted to use vegetative plant materials as their only source of feed. One way or another, these animals rely on a highly active population of micro-organisms in their gastrointestinal tract. The anatomical and physiological adaptations are diverse, with the microbial population in some acting in the hindgut (the caecum and colon) and in others in the foregut (a sacculated stomach in animals like the kangaroo; and a complex compartmentalised forestomach in the ruminant animals such as sheep, goats, cattle and deer).

There are two basic types of digestive systems: ruminant and non-ruminant. Ruminant digestive systems are highly adapted to a diet of grass that has to be extensively chewed (to ruminate means to 'chew again'). The key feature is the process of microbial fermentation of feed that occurs before the normal animal digestive process. The ruminant is typified by a complex stomach with four parts: the rumen, the reticulum (together called the paunch, where micro-organisms live at high concentrations in an anaerobic environment), the omasum, and the abomasum (the equivalent of the non-ruminant stomach). The rumen and reticulum contract frequently to mix the rumen contents and maintain optimum conditions for fermentation, and move contents into the omasum and abomasum.

The non-ruminant class of animal includes animals that may be herbivores (such as the horse or the rabbit) but they do not have the paunch of the ruminant herbivores. They may be single-stomach omnivores such as pigs, which cannot digest grass efficiently. Such animals may have a major fermentation process in the hindgut (caecum and colon) acting on the food residues that remain after the normal animal digestion processes, but cannot take advantage of any microbial growth and synthesis of vitamins unless they reingest their faeces. Poultry have a different kind of complex stomach system, with a crop where enzymes work and a gizzard where food is ground finely.

The process of breaking down foodstuffs into their useable forms is done by enzymes contained in gastric, intestinal and pancreatic secretions in the early stages of transit of digesta through the digestive tract and by bacteria that either precede (in ruminants and macropods) or follow (in all herbivores and some omnivores) secretory digestion. The nutrients derived from the digestive process (including products of fermentation where it occurs) are absorbed through the mucosal wall of the digestive tract and into the bloodstream (in the case of most water-soluble nutrients) or the lymphatic system (in the case of fats and fat-soluble nutrients).

The bloodstream distributes the nutrients to the cells of organs and tissues, with most of the absorbed nutrients taking a first passage through the liver. Making soluble nutrients out of foodstuffs in the ruminant involves extra complexities since both readily digestible substances and those not digestible by secreted enzymes are susceptible to microbial fermentation. Thus not only sugars and starches (readily digestible) but also cellulose and other complex carbohydrates (not digested by secreted enzymes) are broken down into short-chain fatty acids and some other organic acids, which are absorbed from the rumen, providing the animal with energy-yielding substrates.

Dietary proteins are broken down into amino acids and ammonia; these, and importantly ammonia derived from non-protein nitrogen compounds, are in part utilised by the micro-organisms to synthesise their cell proteins. The microbes also synthesise vitamins B and K. The animal reaps a nutritional benefit when it subsequently digests the microbial cells after they flow to the intestines. Ruminants also absorb ammonia, which is subsequently used to synthesise urea in the liver. This is either filtered by the kidneys and excreted in the urine or is recycled back to the rumen through saliva. In this way ruminants have evolved to conserve nitrogen.

Non-ruminant animals such as pigs can be fed controlled diets. These are strictly specified compounded feeds that provide the different balances of nutrients to support, for example, high biological efficiency of reproduction, rapid pre-weaning growth of offspring and high efficiency of post-weaning growth. Grazing ruminants eating pasture choose their own feed. Extra economic benefits may arise through supplementary feeding of ruminants.

Reproduction

Animals have a sexual reproduction system in which ova, produced and released by the female's ovaries (ovulation), and spermatozoa, produced by the testes of the male, join to form a single cell with genetic material from each parent. The female animal will mate with the male only at certain intervals and for some species and breeds of livestock, at restricted seasons of the year (seasonal breeders). The time when the female will not mate is termed *anoestrus* (infertile months). In the seasons when mating will occur there is a distinct cycle (the *oestrus* cycle) that involves a short period of receptiveness (oestrus, when the female is often described as being 'on heat'). Ovulation occurs at a time closely associated with oestrus.

Different types of animals typically shed different numbers of ova: a cow sheds one or two, a ewe sheds two or three and a sow sheds up to 20. For animals that are seasonal breeders the numbers of ova shed (ovulation rate) varies. Greater numbers are shed in the mid to late stage of the breeding season, and with high or improving nutritional status. Physiological and behavioural mechanisms are involved in the full suite of reproductive processes.

The animal's reproductive processes are controlled by a complex set of peptide and steroid hormones; key hormones are produced by endocrine glands of the reproductive axis, the hypothalamus, the pituitary, the ovaries and in pregnant animals the utero-placental unit. The main regulatory gland for reproduction is the posterior pituitary gland found at the base of the brain.

Once fertilisation has taken place, the fertilised ovum, called a zygote, starts growing by the process of cell division. As this proceeds, the embryo moves down the fallopian tube and attaches to the wall of the uterus (implantation). The timing of these events varies considerably between species. Next, cell differentiation occurs, where cells of the embryo start to form different components of the animal-to-be and the placenta is formed. This becomes the conduit for nutrients from the mother to the foetus and for the elimination of waste products made by the foetus.

At about 10 per cent of the gestation period the tissues and organs are differentiated, and the foetus develops and grows until birth. At the end of the gestation period, hormonal changes cause, or make possible, the separation of the foetus and the placenta from the uterus. The offspring is born, and immediately needs to breathe oxygen. Metabolic functions have to be served by substances taken in from the external environment. Initially the mother provides milk.

Reproductive performance can be measured by the number of offspring born live or the number of offspring surviving beyond a critical period after birth or up to weaning. The last of these is a summation of the processes affecting fertility, fecundity and survival. It is the product of the ovulation rate, fertilisation rate, survival rates of the fertilised ovum as embryo, foetus and newborn, and subsequent survival while nurtured by the mother. The efficiency of reproduction in farm animals is assessed in terms of number of offspring born alive, or number weaned, per female available to be mated, per year. Thus if a flock of 1000 Merino ewes has 800 lambs born alive, this is a live lambing percentage of 80 per cent. If 750 of these lambs are alive at weaning, weaning percentage is 75 per cent.

In intensive livestock production systems with non-ruminant species, where the physical and nutritional environment can be tightly controlled, conditions for reproduction are controlled at all stages to achieve close to the full potential weaning percentage for the mated females. Pigs are managed in this way.

In grazing livestock production systems most aspects of the environment are poorly controlled and a primary management consideration is timing of mating to optimise effects of variables such as seasonality of the animal and the seasonal pattern of nutrition. Feed supply has to suit various key stages in the sequence from ovulation to weaning and beyond, to conditions for post-weaning growth to attain target weights. Nutrition is the key to good reproductive performance in grazing animals.

Some species of grazing animal have a distinct breeding season. Thus British breeds of sheep have their oestrus cycle during autumn and winter and are anoestrus in spring. Most Merinos will mate from late spring to the end of winter, though some strains of Merinos will mate right through the year. Cattle do not have a distinct breeding season and are willing to mate right throughout the year. Breeding seasons occur because of hormonal responses to changes in day length. Shortening day length, such as occurs from late December until late June, stimulates breeding activity in ewes whereas the lengthening days from late June to late December inhibit breeding activity.

Reproductive performance is a key factor in determining the range of production alternatives since it sets constraints on the herd or flock structure, the replacement policy, the selection pressure and the rate of genetic improvement.

Lactation

Milk is produced in the mammary glands of the female animal that has given birth. The growth of the mammary glands and the production of milk are controlled by hormones. The milk produced by the female in the first three or four days following birth is called *colostrum* and contains antibodies that destroy invading bacteria, thereby helping to prevent diseases of the new-born animal. Normal milk meets all the young animal's nutritional requirements at first, then a mix of milk and solids are consumed until the animal is weaned and is independent of mother's milk. If milk is the primary commercial product, as with dairy cows and milking goats and sheep, the offspring will be removed and managed separately.

The onset of lactation follows parturition. Lactation is stimulated by the hormones prolactin and ACTH. These are released by the anterior pituitary gland. During pregnancy, milk-producing units of the udder, called alveoli, grow. Alveoli are the simplest milk secretory unit. They are lined by secretory epithelial cells that produce milk, which is then secreted into the cavity of the alveoli. In the udder the alveoli open into small ducts that join to form large ducts, which drain into the cistern and teat cavity. Milk production after calving is partly determined by the amount of alveolar growth. This occurs mostly in the latter part of pregnancy. Adequate nutrition is needed for the alveoli to grow. Nutrition in the latter part of pregnancy is an important determinant of milk production after calving.

The release of milk from the alveoli in the udder, called 'milk let-down', is caused by the contraction of cells surrounding the alveoli, called myoepithelial cells. This contraction occurs in response to the release of the hormone oxytocin from the posterior pituitary. The release of oxytocin is a reflex response to stimulation of the teats, such as the sucking of the calf or washing the teats with water. The release of oxytocin, and milk let-down is blocked by the release of adrenalin. Frightening dairy cows causes the release of adrenalin and prevents them from letting down their milk.

Growth and development

Growth is the increase in size of the cell, tissue or organism. Development is the progressive specialisation of cells, both structurally and functionally. The major growth phases of mammals are:

- *prenatal* (embryonic, foetal), influenced markedly by maternal and placental factors, and programmed to produce a viable new-born animal
- *early post-natal* (suckled, infantile, prepubertal, juvenile), influenced by maternal lactation level, and nutritional and health conditions following weaning
- *maturing* (adolescent, postpubertal, to a predetermined mature body size), influenced by nutrition, health and, in the female, the competing demands of pregnancy and lactation.

Growth is typified by changes in dimensions, composition and function. Body tissues develop at different rates usually in the following order from first to last: brain and nervous tissue, bone, muscle and fat. Body composition refers to the relative proportions of muscle, bone and fat. It can also be represented chemically as the proportion of water, protein, fat and minerals. Baby animals are mostly bone and muscle and as they age the content of bone, muscle and fat increase at different rates. Over time, size, shape, weight, skin and coat of the young animal change as growth (increase in linear dimensions and weight) and development (changes in body composition and tissue function) occurs.

In the first half to two-thirds of the gestation period, growth of the placenta dominates. This may set constraints on the sufficiency of transfer of nutrients from maternal circulation to the foetus. Rapid growth of the foetus occurs in the last third of pregnancy. Birth weight and survival prospects of the offspring depend on nutritional adequacy during the placental growth phase. Placental development affects capacity to transport nutrients in the last trimester of pregnancy, when nutrient demands of the growing foetus are considerable. Growth rate increases further following birth, subject to milk production of the mother and any check that occurs at weaning, until after sexual maturity is reached. Thereafter growth rate slows.

Rapid growth means the animal is converting food into live weight at a high rate. This is usually associated with a higher proportion of growth of lean tissue (e.g. muscle); 1 gram of protein deposited is equivalent to 5 grams of weight gain because lean tissue consists of 25 per cent protein and 75 per cent water. This comes from high intake of nutrients, the balance of nutrients in the diet, or high efficiency of metabolism by the animal. After sexual maturity, the rate of growth slows. The feed conversion ratio (feed intake per unit of liveweight gain) is higher than when the animal was younger. In part this is due to the tendency for fat to be an increasing component of liveweight gain as animals grow older, and deposition of fat has little water associated with it. Feed shortages cause an animal's rate of growth to fall below the rate that would have occurred under conditions of adequate nutrition. When adequate feed is subsequently supplied (often referred to as realimentation), an accelerated rate of growth, known as catch-up or compensatory growth, will occur. The extent of 'catch-up' growth depends on numerous factors, including the severity and duration of the period of inadequate nutrition, stage of growth of the animal, and the species.

Environmental adaptation

Animals and their performance are affected by feed value, air temperature, humidity, rainfall, solar radiation and health status.

Weather

The effort an animal must make to maintain normal body temperature when ambient temperature varies, or it is chilled by rain and wind, affects the rate the animal is able to grow, reproduce, and make milk. For instance, in a cold environment, food energy and body energy reserves are used to increase metabolic rate to produce body heat at the expense of productive functions. Thus, young animals, particularly newborn, may not survive a period of chill or dehydration. Some individuals or types of animals adapt better to particular environmental conditions than others, because of body size, body fat stores, or coat type. In some instances it has proven economically advantageous to protect the production animals from variable climatic conditions by housing them. Holding animals in constructed environments means confining them to a restricted space, and this brings into play other management problems.

Disease

Healthy animals produce more efficiently, and disease is often a signal of management problems. Disease may be stark (clinical), the signs being obvious disabilities or lesions; or they may be perceptible only through slow growth and low productivity (sub-clinical). Diseases are classified by cause: bacterial, viral, fungal, internal and external parasitic and metabolic. Australia is fortunate in not having some of the most devastating animal viral diseases, such as foot and mouth disease, scrapie, and bluetongue. This good fortune has been protected so far by rigorous quarantine regulations. Animals are protected by vaccination from some important bacterial diseases such as the clostridial diseases known as enterotoxaemia, pulpy kidney, blackleg and tetanus.

The diseases suffered by grazing livestock are different in different parts of Australia. For instance, in the tropical areas cattle tick, tick fever and buffalo fly are major problems. In the high-rainfall areas of the south major concerns are with internal parasites (helminths, liver fluke), external parasites such as blowflies, weaner ill-thrift due to worm infestation (among other things), mastitis in dairy cattle and footrot in sheep. Other diseases include vibriosis, which can cause infertility and abortion in cattle, and disorders due to nutritional circumstances such as white-muscle disease, a selenium deficiency affecting young lambs, bloat in cattle, grass tetany in cattle and sheep, milk fever and ketosis in dairy cattle, and ryegrass toxicity in sheep. Disease organisms respond to particular climatic circumstances, and the risk of infection increases with increased animal density. Disease becomes a problem when nutritional stress is placed on the animal. Poor condition predisposes an animal to infection and can also be a result of an increase in an existing low-level infection.

It is possible to control the major animal disease problems by using vaccines and by having good hygiene and husbandry. Poor animal health has a high cost in terms of treatment costs and losses in production. Effectively implementing preventive management practices, accurately diagnosing and vigilantly monitoring animals are the management means to control disease.

Feed conversion

In animals, all tissue components are in a dynamic state, continually being degraded and replaced. This process of turnover is energy consuming and leads to net losses, no change, or net gain in amounts of the respective components. Products of degradation may be reutilised, but there is always a proportion that is swept into metabolic pathways, leading to oxidation. Replacing these tissues, and the inputs of energy needed to drive the metabolic transactions involved, is ultimately provided by the intake of nutrients. Feed ingested by the animal will, with varying effectiveness, supply substrates contributing to energy supply, and to synthesis of protein, lipids and other cell constituents.

Protein provides amino acids and small peptides for absorption. Amino acids that the animal tissues cannot produce through metabolic conversions are particularly important as they must be provided in the feed supply. Vitamins and many minerals act as metabolic cofactors in these processes. Animals transform energy, and synthesise proteins and lipids at rates that support maintenance (no change in weight or physiological state), growth, pregnancy and lactation.

An animal's energy requirements are defined in terms of the amount of energy needed for maintenance for any defined environment, the quality of feed and any work or exercise requirement. Energy required for maintenance is determined by size and age of the animal. Additional to maintenance energy requirements are energy needed for weight gain, pregnancy and lactation. An animal's need for energy above the maintenance requirement increases as the animal expends energy grazing the feed and then digesting it. Also, as pasture quality deteriorates, more energy is needed for chewing, mixing digesta and transferring it, and for digestion. Exercise too is important. In feedlots, the exercise factor is approximately 15 per cent above maintenance; when grazing medium to good-quality pastures this increases to 35 per cent. On poor sparse pastures it rises as high as 50 per cent above maintenance. Disease increases energy requirements, as also does psychological stress. High and low temperatures and wet windy conditions raise energy needs. Inadequate feed causes slow or retarded growth of young animals, especially females, and leads to a delayed breeding performance and often lower lifetime productivity. Supplementary feed to help achieve good early growth rates is profitable when viewed from the lifetime performance of the animal.

Australia's grazing animals are ruminants. The high populations of microbes in the rumen are able to convert cellulose to absorbable, energy-yielding fermentation products, and make essential amino acids. The way a ruminant converts intakes of grass and other herbage into utilisable energy-yielding substrates and amino acids, and eventually to valuable animal tissue and product, is a complex process, especially when compared with the way single-stomach animals operate. The feature of ruminant nutrition is that the composition of the foodstuffs eaten is different to the composition of the supply of nutrients to the animal. The main practical implication of the ruminant methods of digestion is that the relationship between feed intake and eventual effect on body weight or animal product produced is imprecisely known. The animal chooses its own diet from a range of types of herbage available in the paddock, and the feed conversion process is lengthy and complex. It is not easy to identify precisely the relationship between feed eaten and the growth and the animal product that results.

The amount of a foodstuff an animal consumes is partly determined by how rapidly the organic matter of the feed can be processed by the rumen and removed from the rumen. The ease of digestion of a feed partly determines how much feed an animal will eat. Intake of pasture and digestibility of pasture are highly correlated. The amount an animal consumes is determined by digestibility and by the balance of nutrients it gets from the feed. Anything that increases the rate of digestibility of a feed leads to increased feed intake and increased production of a ruminant, providing the balance of nutrients is adequate and critical nutrients are not deficient or limiting. Palatability, as well as environmental factors such as temperature and the spatial distribution of the forage, all affect how much feed an animal is able, or wants, to eat. Short-term regulation of feed intake in animals involves a complex interplay between the physical components of the feed consumed, products of digestion and fermentation, metabolism of absorbed nutrients and how these influence the satiety and appetite centres of the brain. This process is modified by changes in animal physiological state and learned behaviours (preferences and aversions).

While some animals are genetically better than others at converting feed into bodyweight, stage of growth also has a big influence on the efficiency of feed conversion and the profitability of a feeding operation. The higher the feed conversion ratio (FCR), the less efficient is the feed conversion process. The stage of growth of animals has a direct impact on how efficiently they convert feed into liveweight. The older an animal is, the greater the amount of feed needed for each kilogram of weight gain, generally because they deposit more fat than lean tissue.

The efficiency of the conversion of feed to output by the animal is enhanced by increasing the availability of essential amino acids to the animal. This is done by increasing the efficiency of microbial growth in the rumen; increasing the amount of dietary protein that escapes rumen fermentation and is digested in the intestines; increasing the digestibility of fibrous feed in the rumen, which increases both amino acids and volatile fatty acid production; and by decreasing the use of amino acid for other metabolic functions. The effects of any of these steps are enhanced if the animal's feed intake is stimulated.

Feed requirements

The feed requirements of animals vary through the year, depending on whether the animal is non-pregnant, in early and late pregnancy, lactating, growing or gaining weight. A common practice is simply to compare the total feed needs of various classes of animals in terms of maintenance feed needs of a standard livestock unit, such as a 48 kg bodyweight grazing wether for some specified time span. This unit is termed the dry sheep equivalent (DSE). The DSE is specified in terms of the maintenance feed requirements of a 48 kg wether for a year. This is 8–9 megajoules (Mj) of metabolisable energy (ME) per day for a wether grazing medium-quality pasture, which translates into around 3000 Mj of ME per year. At 10 Mj of ME per kg of dry matter (DM), 300 kg of DM would maintain a wether for a year. A crossbred lambing ewe may be rated as having an annual feed requirement equal to that of 1.66 DSE; a 320 kg steer may be rated as equivalent to 7 DSE, and so on. In predominantly dairy, beef cattle or fat lamb areas, the relevant measures may be the milking cow unit, breeding beef cow and ewe unit respectively.

Standard animal feed ratings are based on total feed needed over the whole year, rather than the demand at a particular time. For management, feed demand of animals at different stages during the year is important. Refinements of annual **stock equivalents** such as the DSE for assessing feed demand and supply emphasises the energy needed by animals and the energy supplied by feeds through the year, on a monthly or seasonal basis.

Metabolisable energy

The common measure of the feed requirements of an animal is the amount of metabolisable energy it requires. Metabolisable energy is measured in megajoules and is defined as shown in figure 6.2.

FIGURE 6.2 Partition of food energy

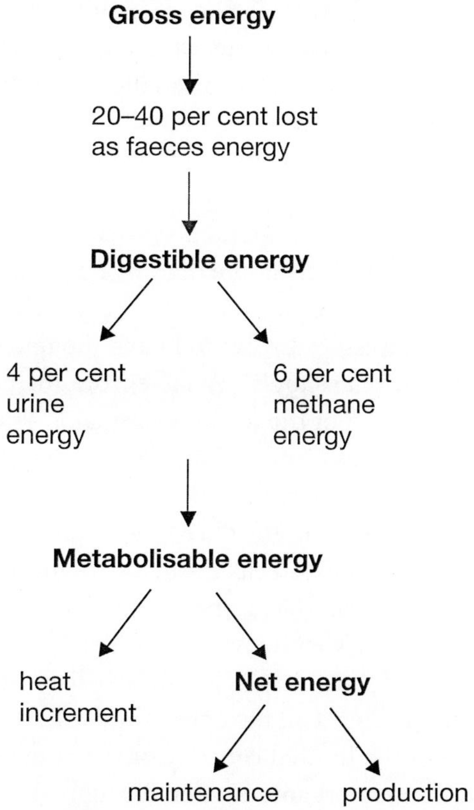

As the energy required by animals varies depending on the quality of the feed they are consuming, precise energy requirements of animals need to be specified in terms of particular feed qualities. Various equations are available to estimate the amount and type of energy requirements of animals for different stages of their production life under Australian conditions. However, such estimates give little recognition to the fact that energy can differ in its uses in the animal, and that energy supplied will differ according to the combination of feeds supplied to the animal. Furthermore, the constant genetic improvement of livestock for a range of production traits means that the various equations used to estimate energy requirements may be less reliable over time.

Pastures and grazing

The nutrient composition of available pasture changes in a matter of days, and from season to season, and total feed intake and the selection of different pasture species to eat is a decision made by the animal. In addition, an animal's requirements for a range of nutrients (amino acids, glucose and liquids) varies according to the physiological state of the animal, and the environmental conditions of heat or cold. All of these factors make it difficult to assess and provide the feed requirements of an animal so as to achieve good performance. Not only does the animal require different dietary ingredients for different production tasks, but when one ingredient is missing or inadequate and the animal has to tackle a particular task of production in a slightly different way, this has effects on the type and quantity of nutrients that are available for the animal's other production tasks. That is, there are interactions between the requirements for and supply of nutrient components in the digestive processes.

The nutritive value of a pasture plant varies as the plant becomes increasingly fibrous over its life. It is necessary to aim to match the nutritional demands of the animal with the nutritional supply available from the pasture. Nutritional supply depends on how much of the feed the animal is able to consume. This in turn depends on how fibrous is the herbage. As low-fibre herbage has higher nutritive value than high-fibre herbage, the focus in grazing management is on ensuring sufficient low- to medium-fibre forages are available to animals at the times when they need it most. Peak animal demand for nutrition, at late pregnancy and from parturition to weaning, is timed to occur at the time of the year when plant growth is leafy with low fibre content.

At times in their production cycle animals will be unable to consume enough high-fibre herbage to get all the energy and other nutrients they need. This happens because of the limited capacity of the rumen and the slow rate of passage of highly fibrous organic matter through the rumen. There is a critical balance for ruminants between consuming sufficient fibre to maintain a healthy rumen and consuming enough feed to provide nutrients for productive purposes.

The duration of the supply of high-quality low- to medium-fibre herbage is determined partly by the length of growing season and related climatic factors. The production of milk and meat, and the running of breeding activities are most successful where low- and medium-fibre forages make up most of the animal diet. The timing of lambing or calving, fattening and turn-off is in turn largely determined by when the highly nutritive, low-fibre forages are available. At the other end of the spectrum, a wether flock for wool production can be run on a feed supply that contains little or no low-fibre herbage, such as happens in the semi-arid pastoral zone.

Feed supply is the major but not the only factor determining 'what-happens-when' in a grazing activity. Other considerations can mean the ideal time for lambing, calving, fattening and sale of young stock, in terms of feed supply alone, is not the ideal time in terms of the market (if everyone sells at the same time, price will be lower than at other times of the year); or in terms of animal health factors such as female animals not ovulating at certain times of the year (for example as happens with British breed ewes in spring); or in terms of competing demands for labour at certain times of the year.

Seasonality of pasture production, both within and between years, is the major limitation to animal production in temperate areas. The total amount of feed produced over a year for any given fertiliser, stock, soil and plant species, is broadly a function of rainfall and range and duration of temperatures. As such, it varies greatly from year to year. The pattern of growth of pasture plants over a year is usually expressed as kilograms of dry matter (DM) per day, and these rates reflect the availability of moisture and temperature. Changes in the digestibility of dry matter through the year also occur. These changes are more marked in tropical species and areas than in the temperate species and areas. Digestibility of tropical grasses is generally less than for temperate grasses, and the change in digestibility of tropical grasses through the year is far greater than is the case for temperate grasses.

Temperate plants are able to convert light to herbage DM at rates of up to 3–5 per cent of the photosynthetically active radiation. Rainfed pastures in southern Australia produce 6 to 12 tonnes of DM per hectare per year. In Mediterranean climates, sown annual pastures produce high-quality herbage during the mild wet winter and spring, following the autumn rains. In the temperate regions of Australia the rainfall is distributed evenly though not reliably through the year. The moisture regime is adequate for perennial species such as phalaris, rye grass and cocksfoot, to survive. In the northern sub-tropical and tropical areas, rainfall promotes summer growth, and winter pastures are generally not of high quality.

Stocking rate

The stocking rate of an animal activity refers to the number of animals grazing on a specified area for a specified time and using a specified amount of supplementary feed in the most likely seasonal conditions. Different numbers of the same types of animals grazing on an area of pasture over a specific time will be able to harvest different amounts of feed per head simply because of the different ratio of animal:grass feed. Accordingly they will have different rates of growth of body weight or animal product. At a low stocking rate, production per animal is limited by voluntary intake. At a high stocking rate the amount of pasture present will decrease until it restricts animal intake. At the high stocking rate, defoliation stimulates new growth. So different numbers of the same types of animals grazing an area of pasture for a specified time will be confronted with different amounts of grass per head, not just because of the different number of animals on the area of pasture, but because different numbers of animals grazing causes the composition and growth of the pasture to change. Stocking rate is a function of pasture production and pasture production is a function of stocking rate.

As stocking rate increases there will eventually be a decline in production per head from the area of pasture. Generally the producer is more interested in total production per hectare, as amount of feed (land, area and pasture production per hectare) is usually limited relative to the number of animals that the farmer could run on the farm with the labour and capital available. Hence, the farmer is more interested in kilograms of wool or meat or litres of milk per hectare than in wool cut per head of sheep, or carcass weight per lamb or vealer or litres of milk per cow.

Implicit in much discussion about increases in stocking rate is the idea that the extra animals and the pre-existing animals consume the same proportion of available grass after the increase in numbers as they did before the increase. In practice, as stocking rate increases so too will percentage of pasture utilisation. To further complicate matters, the pasture composition itself will change as stocking rate increases. Thus the eventual change in animal performance per head over the whole flock or herd depends on both the number of extra animals, the changes in pasture utilisation by the animals, and the change induced in the pasture itself.

The key point is that pasture helps determine animal production and animal production helps determine pasture production. However, as pasture production itself responds positively to increased stock per hectare, over a range it is often possible to increase stock per hectare without any marked decrease in production per head of all the pre-existing animals grazing the pasture.

Physical production per hectare does not indicate the best stocking rate. This can only be determined by looking at the costs and returns associated with each different stocking rate. Increasing stocking rate increases risk, and increased risk increases costs. The higher the stocking rates the sooner feed shortages and drought arrive. Conversely, low stocking rates greatly increase the likelihood of potential feed supply not being achieved as pasture quality deteriorates more rapidly and total feed supply is less. Still, more pastures are ruined by understocking than by overstocking.

Animals that lose weight in the minimum feed period may compensate by growing more rapidly in the feed surplus period than would have been the case if they were fed better in the minimum feed period. This is called compensatory growth. Or, an animal provided with extra feed such as dairy cows fed in the bail may respond by substituting the supplementary feed for grazing pasture, resulting in less utilisation of pasture and thus causing deterioration in pasture quality.

The farm manager has to allow for variability of feed supply. Calving, lambing, lactation, and buying and selling times are planned so that animal demand matches, as closely as possible, expected feed supply. This will mean that the period of minimum expected feed demand matches the period of minimum expected feed supply. Feed supply in any year will be more or less than the average that is expected to occur over a run of years.

Feed deficits can be handled in several ways. Conserving fodder in periods of feed surplus for feeding in periods of feed deficit is one tactic. The value of conserving feed depends mainly on the length and frequency of bad seasons, but also on the cost of conserving and storing feed, and the price to which feed rises during poor seasons. Buying fodder or concentrates such as cereal grains is usually efficient. A useful strategy is to have a flexible stocking policy that allows the sale or purchase of a buffer group of animals (dry ewes, wethers, dry heifers, steers) according to seasonal feed supply. The conventional wisdom for managing drought feed shortages is to 'sell, prepare to repent, but sell early'.

When the feed supply in a given time exceeds the expected quantity required by the animals it can be used in the following ways:

- fed to existing stock to improve bodyweight
- fed to extra stock
- conserved
- carried over into the following period, with some loss of feed quality.

Animal replacement

Ageing or non-productive animals in the production nucleus of the herd or flock, or young animals turned off for market, need to be replaced. Replacement of animals is by purchase or by breeding. Animal production activities can be categorised into three types: non-breeding; breeding with bought-in female replacements; and self-replacing by breeding with females bred within the system.

Non-breeding animal systems include wethers for wool, fattening store lambs, growing out store heifers, finishing grower pigs and broiler production.

Breeding with bought-in female replacements includes buying in first-cross ewes, Merino ewes, joined heifers, older joined cows with calves at foot, gilts, or sows and litters.

Self-replacing animal systems include straight-breeding Merino × Merino, Hereford × Hereford, or a cross-breeding system such as in a self-replacing Merino ewe flock where a major proportion of the ewe progeny are mated to Merinos, but some to Border Leicester rams to produce first-cross offspring.

The widespread practice is for animal activities to be self-replacing. In this case, as animal production performance tends to increase then decrease over time, the critical factor is the age at which to cull animals. The strategies based on breeding depend on reproductive rate, and this directly affects the possible alternatives in herd or flock structure (i.e. the proportions of animals of different sexes and ages).

A flow chart is a sketch of what will happen in the animal production system over the coming planning period. It shows which animals go where within the animal system, in what numbers, and when it happens. Rates of birth, culling, death and **cast for age** (CFA) figures, are used to calculate the numbers of each class of animal within each activity that are on hand at various times, and which ones will be sold and retained.

In general the replacement decision in a breeding ewe flock involves:

- to breed or to buy in
- if buy in, at what age
- at what age to cull ewes.

The answer to these questions depends on the price that can be obtained for the cull animal, the expected future production if the animal is not culled, the value of the animal's products, and tax aspects.

Rarely in practice does an animal activity have a stable size and composition over time. At any time that a livestock enterprise is analysed it is more likely to be changing in size or composition than to be routinely turning off and replacing stable numbers. This is because of changing product prices and seasonal conditions faced by farmers, and their opportunistic responses to them. Another situation is where a grazier has a flock of the one age group, and so production and replacements might be low at first and higher later.

A change in replacement policy affects several aspects of an animal activity. In a breeding flock the number of animals that can be carried through the most limiting feed period is restricted by the number of younger stock also having to be fed. The number of young animals depends on the number of ewes in the breeding flock. A change in replacement policy has ramifications for the feed available for all of the activities on the farm, and for whole farm profitability. Taxation considerations are also important in decisions about replacement.

Genetic gain

Genes occur in all cells of an animal's body on chromosomes. The chromosomes are in pairs and one of each pair comes from the male and one from the female. Animal genes control the animal's biological processes. The characteristics of any animal depend to a large extent on the genes it receives from its parents. Within any animal population there is a variation in individual performance expressed, for example, in different milk production, growth rate, wool cut, carcass composition, lambing rate. Animal breeders can use this variation to gain genetic improvement by keeping, for future breeding, those animals that have the highest levels of performance. Only some of the better performance of such animals is due to a superior collection of genes; the rest is due to a better environment and/or a positive interaction between the environment and the expression of superior genes. The annual gain in physical terms of any character under selection in a breeding program can be estimated as follows:

$$\text{Annual gain} = \frac{(\text{Selection differential} \times \text{Heritability}) - (\text{Inbreeding depression})}{\text{Generation interval}}$$

where:

- *selection differential* (SD) is the difference between the observed performance of selected animals and the average performance of the herd or flock to which they are being mated. If the flock averages a wool cut of 4.5 kg, and the cut of selected individuals averages (after age and sex correction) 6.0 kg, then the selection differential is 1.5 kg.
- *heritability* (H) is the extent to which differences between animals are transmitted from one generation to the next, expressed as a decimal or a percentage. When using published heritability figures for budgetary calculations, it is important to ensure that the environment in which the heritability figures were derived is fairly similar to that in which the breeding program will be carried out.
- *generation interval* (GI) is the length of time between one generation and the next. The average age of the parents is a measure of the generation interval. The longer the genetic interval, the slower is the annual rate of genetic progress.
- *inbreeding depression* (ID), or inbreeding percentage per generation, is estimated as $100/8 \times S$, where S is the number of sires used. For example,

in a closed herd, six sires would achieve an inbreeding percentage per generation of about 2 per cent.

$$\text{Gain per generation} = SD_{(\text{average of sires and dams})} \times H$$

$$\text{Gain per year} = \frac{SD_{\text{sires}} + SD_{\text{dams}} \times H}{2GI}$$

For example, if the average SD of sires and dams is 1 kg, the H of wool weight is 0.3, and the GI is 4 years, and there is no inbreeding depression, then the annual gains is:

$$(1.0 \times 0.3) \div 4 = 0.075 \text{ kg}$$

The value of the extra output is the annual gain in output multiplied by the value per unit of output minus the extra costs associated with the extra output.

The requirements for getting good results from a breeding program are:

- good-quality foundation stock
- a high selection differential, that is, clear superiority of selected individuals over the average
- testing under uniform environmental conditions
- a short generation interval
- a low inbreeding rate.

In terms of the equations for genetic gain, the requirements for fast gains are:

- maximise SD and H
- minimise GI and I.

REVIEW QUESTIONS

1 What are the determinants of an animal's feed conversion ratio?
2 Describe the typical growth and development phases of a farm animal.
3 Stocking rate is a function of feed supply, and feed supply is a function of stocking rate. How?
4 Discuss the main differences between ruminants and non-ruminants with respect to digestive processes.
5 Explain how metabolisable energy is derived from gross energy intake in an animal.

6 Stock equivalents are crude measures of animal feed demand. Explain what stock equivalents are and how they are used. How does the metabolisable energy system relate to the use of stock equivalents?
7 Food intake is one of the key determinants of the performance and productivity of grazing ruminants. What are some of the factors that can influence intake in grazing ruminants? Why is it difficult to predict intake in a grazing situation?
8 Describe two ways to measure reproductive performance in breeding ewes. Why is adequate nutrition important for optimal reproductive performance?
9 Understanding basic animal processes is important for improving livestock management. Discuss.

FURTHER READING

Berg, R. T. and Butterfield, R. M., *New Concepts of Cattle Growth*, Sydney University Press, 1976.

Brightling, A., *Livestock Diseases in Australia*, C. H. Jerram, Mount Waverley, 2006.

Butterfield, R. M., *New Concepts of Sheep Growth*, Department of Veterinary Anatomy, University of Sydney, 1988.

Cottle, D. J. (ed.), *Australian Sheep and Wool Handbook*, Inkata Press, Melbourne, 1991.

CSIRO, *Nutrient Requirements of Domesticated Ruminants*, CSIRO Publishing, 2007.

Freer, M. and Dove, H., *Sheep Nutrition*, CSIRO Publishing/CABI, 2002.

Lindsay, D. R. and Pearce, D. T. (eds), *Reproduction in Sheep*, Cambridge University Press, 1984.

Morley, F. H. W. (ed.), *Grazing Animals*, World Animal Science B, Elsevier, Amsterdam, 1981.

WEBSITES

Australasian Farm Business Management Network: www.csu.edu.au/faculty/science/saws/afbmnetwork/

CSIRO: www.csiro.au/

National Farmers Federation: www.nff.org.au/

NSW Department of Primary Industry: www.dpi.nsw.gov.au/

Vic Department of Primary Industry: www.dpi.vic.gov.au/dpi/index.htm

7 Practice of Animal Production

CONTENTS

Introduction

The aim of operating an animal production system is to make profit from the animal system and the whole farm system. Profit is determined by the large range of factors that combine to place animal product in the hands of the consumers. In the following sections extensive grazing beef, wool and lamb production systems are discussed, followed by intensive grazing dairying, and the intensive controlled-environment pig and poultry production systems.

Beef

Beef is meat from cattle. A carcass is what is left from an animal after removing the head, feet, hide and internal viscera except the kidneys, edible muscle, connective tissue, fat and bone. Fat is the component that varies the most between animals, and the amount of fat to meat (too little or too much) can have a large bearing on the value a consumer puts on a cut of meat. Apart from tenderness, other important traits include meat colour (consumers prefer cherry red meat colour to dark red or brownish colour), fat colour and flavour (both depend on the market into which the meat and cuts are sold). Connective tissue, which supports the muscle and helps hold the carcass together, is strong and tough material such as tendon and ligament, and is the major factor in meat toughness. The value of a cut of meat also depends on the relative proportion of connective tissue to meat, as this determines tenderness or toughness, and tenderness or toughness determines the type of cooking that is required. Cuts with lots of tough connective tissue and muscle fibre require boiling or steaming to soften the collagen of connective tissue into softer gelatin. These cuts of meat are the least valuable. The higher-valued cuts of meat have relatively less of the tough connective tissue and muscle fibre, and can be roasted or grilled.

A carcass is broken up into portions of meat with similar characteristics, and hence, into pieces with distinct uses and value. While the consumer wants particular, relatively small pieces of a carcass, the retail butcher will often buy a carcass or large portions of it, the wholesale butcher buys either whole carcasses or live animals to convert into carcasses, and the farmer produces live animals.

The ratio of the weight of the cold carcass to the weight of the live animal is called the dressing percentage. Cold carcass weight is about 98 per cent of hot carcass weight. Dressing percentage ranges from 65 per cent in well-grown animals with a lot of fat to 45 per cent in poor-quality cattle with almost no fat. Good British breed carcasses will dress out at around 60 per cent, while reasonable animals grown steadily on pasture will dress out at 55 per cent. High dressing percentages are not necessarily a good guide to quality or value of animals, as the high ratio of carcass to liveweight may be due to excessive fat or bone on the carcass and thus not be indicative of the amount of meat, which is what the consumer wants most. Charolais cattle usually dress out higher than the British breeds, often producing 65 per cent carcass weight without too much fat.

The objective in producing beef is to grow as much lean meat as possible in a given time. That is, a heavy carcass without 'excess' fat is required to be produced by a young age. There is some relation between fat and meat quality in that there is a minimum amount of fat inevitably on a carcass if the meat is to be of sufficient quantity and quality. There are various types of carcass fat: mainly subcutaneous or outer layer fat, intramuscular or 'between-muscle' fat, and intermuscular or 'within-muscle' fat, called marbling. A notable feature of these different types of fat is that generally they are 'laid-down' on the carcass in reasonably fixed proportions. The rate at which different carcass fat deposits are laid down can vary depending on the plane of nutrition and growth path of the animal. However, the level of subcutaneous fat will be indicative of fat elsewhere in the carcass, at least, once an animal is not very young and reasonably grown. Breed and sex differences also account for some differences in the relative amounts of different types of fat deposited.

Fat is a high-energy substance and it requires considerable nutrition to deposit fat on the body. Putting feed into fat that does not add to the value of the carcass is obviously a waste of feed, which would be better used to produce meat. Different consumers demand different proportions of meat and fat in the cuts they buy, and different characteristics of fat too, such as colour. The notable example of this is in the Japanese beef market where consumers prefer meat with marbling. This is the last fat to be laid down in a carcass and thus requires a high-energy diet for the animals. That is, animals need to be 'finished' on grain to produce the marbling that Japanese beef buyers prefer.

Beef produced in Australia includes pasture and grain-fed animals that reach saleable weights of 300–350 kg liveweight (180–210 kg dressed weight) in less than one year; young animals (vealers) of six to eight months reared on pasture and weighing 250–300 kg liveweight (150–180 kg dressed); one to two-year-old steers, grown up to 400–500 kg liveweight (240–300 kg dressed) on pasture; three-year-old and older bullocks of 600–700 kg liveweight (360–420 kg dressed) grown on pasture; cull cows with a range of 350–500 kg liveweight (210–300 kg dressed); beef from animals in northern Australia, which can take several years gaining weight in the wet season and losing it in the dry season, to reach 300–450 kg liveweight (180–290 kg dressed).

How farmers produce beef

In reasonable grazing areas of southern Australia the major beef cattle activity is breeding and selling store weaners. These are sold for fattening in the most climatically favoured areas. In the high-rainfall areas with seven months or more of good pasture growth, vealers are produced for slaughter. A large proportion of beef in Australia is produced from herds of less than 50 head, run on mixed sheep–cattle, but predominantly sheep properties. The dairy industry also produces beef in the form of cull cows and surplus calves. In recent years there has been a marked increase in beef produced in grain feedlots, many of which are in southern Queensland and northern NSW, close to the cattle- and grain-producing areas of Australia that are closest to the major grain-fed beef market of Japan.

In southern Australia the beef cattle industry is based largely on the *Bos taurus* (British) breeds of Hereford, Angus and Shorthorn. In the southern areas, as well as the British breeds, are found some of the European breeds such as Charolais, Limousin and Simmental, which are bigger animals than the Herefords, Angus and Shorthorn. The Friesian dairy-breed is also used in beef production, usually being cross-bred with British or European beef breeds.

In northern Australia, the *Bos indicus* breeds such as the Brahman and Santa Gertrudis are well adapted to tropical conditions, specifically heat tolerance and parasite resistance. *Bos indicus* breeds and crosses with British breeds make up over two-thirds of cattle in northern Australia. The *B. indicus – B. taurus* cross-breeding improves the heat tolerance and tick resistance of the *B. taurus* breeds, and improves the growth rates, conformation and temperament of the *B. indicus* cattle.

Australia's beef-only properties are found in the pastoral and drier zones, where small numbers of cattle graze on native pastures on vast areas of land. In these production systems, rainfall is restricted to a short season; soil fertility is very low; pasture quality is usually very poor for most of the year; transport costs are high; management input and labour use is minimal; and stocking rates range from one beast to 30–50 hectares.

In southern Australia cattle is not usually the major enterprise on a grazing property. For example, while cattle and calf slaughter represents 80–90 per cent of the gross value of agricultural production for the Northern Territory, cattle production is only 5–15 per cent of the gross value of animal production in NSW, Victoria, SA and WA. Only about 15 per cent of cattle herds in southern Australia are larger than 150 cattle, and less than 10 per cent have more than 200 cows. Thus a significant proportion of beef produced in the southern parts of Australia is produced on farms where the cattle are run as a sideline to the more important activity, usually sheep for prime lambs, or for wool in the high-rainfall zone, or cereals in the wheat–sheep zone. Indeed, this is one reason why the simple straightbred breeding systems dominate in southern Australia instead of the often productive but managerially more complex cross-bred systems. However, in recent years, cross-bred beef production systems have been adopted by some large producers in southern Australia.

Cattle are run on mixed grazing properties usually because there are complementary benefits to be had in the sheep activity. That is, by having cattle, pasture is utilised better because of different grazing habits of sheep and cattle, or cash flows are diversified, or worm control in the sheep activity is enhanced by being able to have a sheep-free area for a time, and the timing and amount of labour required for cattle are different from those of sheep. Thus, having cattle enables greater production of beef and sheep product for a given limited supply of labour than would be the case if sheep only were run on the property.

Stocking rates of 1.5 to 2 breeders per hectare are achieved in the most productive high-rainfall regions with long growing seasons, without much supplementary feeding in usual circumstances. More typically, one breeder per hectare is run on reasonably improved pastures in mixed grazing properties in the southern high-rainfall zone and the better parts of the wheat–sheep zone. Between 85 and 90 per cent calving is achieved, with the majority of heifers calving at two years.

The timing of operations in pasture-based systems of producing beef depends on the seasonal pattern of rainfall and pasture growth. The time when heifers are first mated depends on the time it takes for them to reach the weight necessary for breeding maturity. Rapidly grown heifers reach sufficient weight, and puberty, at 12–15 months. The nine-month gestation requires that cows are mated in summer for spring calving, or are mated in winter for autumn calving.

Feed supply for the lactating cow is critical because the early rate of growth of the cattle is largely determined by the milk produced by the cow. Then feed available to both the lactating cow and the milk-drinking and grass-eating calf is important. The feed available to the weaned calf determines growth rate.

The major diseases are grass tetany, calf scours, abortion, dystocia, worms, lice, fluke, bloat, cancer eye, pink eye and blackleg. Grass tetany (hypomagnesia) is a metabolic disorder arising from low magnesium absorption from the rumen. It results from low magnesium content of the feed intake, or from high potassium and low sodium intake, which affects the absorption of magnesium. Low blood calcium levels resulting from phosphorus deficiency also predisposes cows to hypomagnesia. Old, fat cows losing weight during lactation are most susceptible to grass tetany. An autumn calving cow in southern Australia is susceptible to grass tetany. They gain weight over the spring and lose weight after calving when they are lactating and on relatively poorer feed.

The use of some animal medications has implications for access to markets. Chemical residues from medications given to cattle, either for disease control or to promote growth, can have implications for public health. Animals treated with certain medications are banned from markets in some countries.

Lamb

A lamb is usually defined as a young sheep having milk teeth only, which means they are less than one year old. Young lambs or suckers are defined as being three to six months old and old lambs range from about seven to 15 months. Mutton is meat from adult sheep, and is produced by lamb and wool flocks. Mutton comes from cast-for-age and cull ewes from both lamb and wool flocks, from wethers of all ages, and from lambs that do not grow sufficiently rapidly to be sold as lambs. The latter are generally termed hoggets (or two tooth).

How farmers produce prime lambs

Lamb is produced mostly from British breeds—Dorset, Suffolk, Border Leicester and so on—and from British breeds crossed with Merinos. The British breed ewes have a restricted breeding season of about six to eight months, although cross-breeding with Merinos increases this period to about 11 months. Gestation is five months. In recent years other European breeds, for example Texel (originally from the Netherlands), have been introduced to Australia. These breeds have large mature size and rapid growth rate, good muscling and high fecundity. All these traits are highly desirable in current prime lamb production. In traditional systems, generally one lamb per year is produced. The British-breed sheep and their crosses have higher reproduction rates than the Merino. The introduction of more fecund breeds and an increased emphasis on genetic selection for fecundity (for example the booroola gene in the Merino 'Booroola Merino') has increased ovulation rates. Twinning rates have steadily increased. Marking percentages (the number of lambs marked per 100 ewes present at marking) for the British breeds and cross-breeds range from 90 per cent to 110 per cent. Multiple ovulation occurs in some breeds, and ovulation rate increases as body weight increases.

Providing extra feed with high protein content for two to four weeks prior to mating affects ovulation rate. This positive affect of a rising plane of nutrition on ovulation rate is termed the 'flushing effect'. Ample nutrition is needed in the last month of pregnancy. A ewe carrying twins is especially vulnerable to poor nutrition (pregnancy toxaemia) in late pregnancy. She will deprive herself of energy, diverting all energy to the foetal lambs until the ewe dies. Paradoxically, very fat ewes carrying twins are also predisposed to pregnancy toxaemia. If the ewe survives, the lambs are off to a poor start, having low birth weight and low resistance to cold. The undernourished ewe has poor milk production and often poor mothering ability.

Losses of lambs after birth can be 10–25 per cent. Losses are caused by exposure to cold, wet and windy weather, and by predators. The ewe is vaccinated against toxaemic clostridium diseases such as pulpy kidney and blackleg in late pregnancy. The growing lamb's major disease risk comes usually from internal parasites. They are vaccinated and drenched to protect against these diseases.

The major determinant of lamb growth rate is the milk supply of the ewe. At about four to six weeks, lambs begin to supplement their mother's milk with grass. Under reasonable conditions a lamb will grow to 13–16 kg carcass weight or 24–30 kg liveweight at four to five months of age. Dressing percentage in lambs is around 48 per cent but this is influenced by a number of factors, especially carcass fatness. At around six months, a lamb carcass will weigh 16–20 kg (30 to 38 kg liveweight). Carcasses of older, larger lambs will weigh over 20 kg dressed weight (40 kg liveweight). Fat is the last tissue to develop, and carcasses are graded according to Fat Score, which is determined by measuring the fat depth (in mm) at the 12th rib, 11 cm from the midline (the 'GR' site). Lambs are sold for slaughter by public auction, or a price is negotiated in the paddock, or they are sold according to an agreed price per kilogram and the class of carcass (called selling 'over the hooks').

Nearly all lambs produced in Australia for meat are grown on pasture. Most pasture is introduced pasture species, and some is irrigated pasture. The nature of the lamb producing operation depends on whereabouts in the country it happens. A result of the seasonal patterns of climate is a seasonal pattern of supply of lamb with a peak supply in the spring declining to a minimum supply of lamb in the summer. Prices through the year are highest in the summer and lowest in the spring. There is a price premium for out-of-season lambs because they are difficult and costly to produce.

The timing of mating and lambing is dictated by expected feed supply and by expected market conditions. In the cold, high-rainfall areas such as Tasmania, southern Victoria and the Central and Northern Tablelands of NSW, winter and autumn is the period in which feed is most limiting, since most pasture growth occurs during spring and summer. In the lower-rainfall areas of the temperate zone, pastures grow most during late autumn, if the early autumn rains arrive, and in winter and spring. Late summer and early autumn is therefore the feed-limiting period of the year. Many prime lambs are turned off irrigated annual pastures of annual grasses and clovers, or in some areas, irrigated lucerne. Also, irrigated and rainfall-opportunistic summer crops are used to 'finish off' lambs when the seasonal pastures have finished before the lambs have 'finished'.

The earliest lambs of the year come from the irrigation areas of southern NSW and northern Victoria. Mating occurs in spring, lambs are born in autumn and are ready for sale from May onwards. Next on the market are mid-season lambs, which come from the wheat–sheep areas of NSW, SA, WA and Victoria. Ewes are mated in late spring, lamb in late autumn and are on the market from August onwards. Next to arrive are what are called late lambs, which come from the Southern Tablelands of NSW, southern, central and north-eastern Victoria, the east coast of Tasmania, south-eastern SA and parts of south-western WA. In these areas matings occur in the summer. Lambs are born in early winter and are ready for market mid to late spring and early summer. Finally, summer lambs are produced, particularly in the Central and Northern Tablelands of NSW and in Tasmania. Ewes are mated in late autumn. Lambs are born in spring and are ready for sale from around Christmas onwards.

Lamb is produced by farmers who specialise in prime lamb production and also on farms where lamb production is one of several other activities. The method of producing lamb depends on the farming system involved, which depends on the location, climate, soil types, topography, water supply and so on. Prime lamb specialists use a second-cross breeding method, in order to reap benefits of hybrid vigour. With this method, first-cross ewes from Merino ewes mated to a British breed ram, such as a Border Leicester, are joined to another British breed, such as the Dorset, to produce second-cross prime lambs. All offspring are sold. Second-cross lambs are produced in the higher-rainfall areas. The first-cross breeders are often bought in from the lower-rainfall wheat–sheep or pastoral zones. In these zones, Merinos are mated to a British breed to produce first-cross ewe and wether lambs. These offspring are either all sold as early lambs; or the wether lambs are sold as lambs or grown to hoggets and sold for meat and the ewe lambs are sold as replacement ewes to be used for breeding second-cross prime lambs.

There are numerous alternatives in the breeding mix for prime lambs. Animals derived from Merinos such as Polwarths, Corriedales and Comebacks are sometimes used as prime lamb mothers instead of the Merino and British breed first-cross ewe, as well as 'composite' breeds such as the Coopworth. While the most common terminal sire used is the Poll Dorset and the Dorset Horn, there are also other quick-growing breeds such as the White Suffolk, Texel and the Cheviot.

A specialist prime lamb producer in high-rainfall southern parts of Australia could be mating Merino and Border Leicester ewes to Dorset rams in late summer – early autumn, to lamb five months later in late winter – early spring, to be marked and vaccinated at about two to four weeks of age, drenched at three to four months of age and to be ready for sale at four to five months of age in late spring – early summer. The lambs will have an average liveweight of 30–35 kg and an average carcass weight of 16–18 kg. However, depending on the breeds used, genetic selection programs implemented on-farm and nutritional management strategies for finishing, heavier carcass weights can be achieved at similar ages.

Several alternative production systems related to prime lamb production are growing in popularity. These include dedicated and opportunistic feedlotting of lambs in the finishing phase; dedicated production of 'store' lambs that are then on-sold to pasture-based or feedlot finishers; and development of dual-purpose Merino sheep lines, for example Australian Meat Merino, which aims for both meat and wool markets.

Wool

Wool fibres grow from follicles called *primary* and *secondary follicles*, which develop in the foetus. Primary wool follicles start being formed at about day 50 of the life of the foetus, and by day 100 they are all producing fibres. Secondary fibres start forming at day 90, and are nearly all formed at birth. A new-born lamb's coat is nearly all primary fibre. Secondary follicles have a sebaceous gland that secretes lanoline, which helps protect the fibre and skin from the effects of water. The number of these follicles is affected by nutrition of the ewe before lambing and just after lambing. Poor nutrition in the last six weeks of pregnancy and in the first six weeks of birth can firstly reduce the number of secondary wool follicles that form, and then reduce the numbers that develop fully and produce fibres.

Wool production in the adult sheep, particularly fibre length, is largely determined by food intake. Undernutrition causes the fibre to grow thinly, that is, with low fibre diameter, and be relatively weak, resulting in tender wool that has a break in it. The incidence and severity of tender wool is exacerbated when there is sudden change in nutrition, as occurs in the late summer – early autumn period when there is a transition from low-quality, poorly digestible pasture to highly digestible, green pasture after autumn rains.

The ratio of primary wool follicles and secondary wool follicles is set at the birth of the lamb and varies with different breeds. Merinos have a high ratio of secondary to primary wool follicles, ranging from 15:1 to 30:1 secondary wool follicles to one primary follicle. In no other breed is this ratio above 10:1, except where a Merino cross is involved. The high ratio of secondary to primary wool follicles leads to a greater diversity of fibres for any area of skin. This causes a finer average diameter of fibres than occurs with more primary and less secondary wool follicles. The length and diameter of wool fibres differ even on the same part of the fleece.

Sheep produce three *true fibres* (wool, kemp and hair). Some fibres have a hollow core and are described as medullated, whereas other fibres are solid and are described as non-medullated. Medullated fibres usually have a high diameter. True wool is a solid fibre, with an outer layer called the cuticle and an inner coat called the cortex. Kemp is the short, hollow fibre found on the legs and face of the sheep. Hair is a long, mostly hollow fibre. The main component of wool fibre is keratin, a protein with a high concentration of sulfur.

The most important characteristic of wool is its fibre diameter. This determines end use. Wool is classed as fine, medium and strong, depending on fibre diameter. Fine wool is less than 20 micron, medium wool is 20–23 micron, and strong wool is 24–28 micron. About half of Australia's wool has an average fibre diameter of 20–22 micron. A tiny fraction (2 per cent) is finer than 18 micron. Fine wool fibres are used for apparel while the coarsest wool is used for carpets. Apparel wools are broken down into two categories depending on length. Longer fine wool is called *combing wool* and is processed by a 'worsted' system. Most of Australia's wools are combing wools. Shorter wool is called *carding wool*, which is processed using the woollen processing system.

Within the Merino breed, different strains have different fibre diameter and wool weight characteristics. The Merinos that produce finer wool are smaller, cut less wool and have a shorter staple than Merinos with the stronger wool. Strong-wool sheep are more susceptible to fleece rot and fly strike than are the finer wool strains. Fleece weight and fibre diameter are highly heritable and thus are the focus of flock genetic improvement. Greasy wool contains wax and vegetable matter and soil. Clean wool yield is the percentage of clean wool, which can be obtained from a quantity of greasy wool. This varies between 60 and 70 per cent of the weight of greasy wool.

How farmers produce wool

Sheep are kept for wool production in all of the non-tropical areas of Australia. The type of sheep and the type of production system used depends on where it happens. In the pastoral zone, sheep are run as single activities, grazing native pastures that receive low rainfall. In other areas, sheep are one of several activities such as cereal grain and legume production or beef. In the wheat–sheep zone, sheep are seen to some extent as complementary to the cropping activity, so the number of sheep run in these areas is affected by the price of wool and the price of wheat. In the high-rainfall area, mixed livestock farming means the number of sheep is affected by the price of wool and the price of beef. In the wheat–sheep areas, sheep will graze a legume pasture that is part of the cropping rotation.

In the arid and semi-arid pastoral zones, one sheep will graze 2 to 12 hectares over the year; in the wheat–sheep zone, 2–3 sheep will graze on one hectare over the year; in the high-rainfall zone, 10–15 sheep will graze on one hectare over a year. Wool cut per head is lowest in the heaviest stocked zone and highest in the lowest stocked areas. Wool cut per hectare is highest in the highest stocked areas and lowest in the lowest stocked areas. The highest lamb marking percentages are in the high-rainfall areas (80 per cent). The lowest lamb marking percentage is in the lowest-rainfall zones (65 per cent).

The amount of wool a sheep grows depends on sex, age, genetic make-up and nutrition. Under similar conditions, males cut more wool than females, and older sheep have heavier fleeces than younger ones. Staple length decreases with age. Nutritional requirements of pregnancy and lactation reduce wool production of ewes by 10–25 per cent. Nutrition of the ewe also affects the number of primary and secondary skin follicles formed in the unborn lamb. Poor nutrition after the lamb is born can mean that wool follicles that have formed do not mature and produce wool. In the adult sheep a lack of nutrition reduces fibre length and diameter, and a period of severe nutritional shortage can cause a 'break' in the fibre.

Merino wool comes from flocks in the arid and semi-arid pastoral zones. These flocks are mostly self-replacing. In the wheat–sheep and high-rainfall areas there are self-replacing and brought-in flocks. In self-replacing flocks, young replacement ewes are selected usually at around 15 months of age, just prior to first mating. The number of replacement ewes selected depends on the age distribution of the existing ewe flock, and the age they are culled. Ewes of all ages may be culled for poor performance, or faults, or poor health.

In a flock in which the number of ewes in each age group is the same and where old ewes are culled at five years of age, then each year one-fifth of the flock is culled for age. The same number of young ewes replace them, plus a few extra to allow for deaths and culls that occur during their time in the flock. Culled-for-age ewes from the harsher climatic areas are still good for a year or two of production in the wetter pasture areas, where they will be used to produce first-cross lambs for sale as lambs, or as replacement first-cross breeders for second-cross prime lamb operations.

On properties where Merinos are run primarily for producing wool, mating is timed so that late pregnancy and lambing occurs when there is the most feed available. Merino ewes will breed well between January and June (decreasing day length), are anoestrus between August and October, and will breed between October and December (although less prolifically than in January–July). This characteristic breeding pattern is important in the decision to lamb in spring or autumn, along with considerations of feed supply and weather conditions around lambing time.

The usual ratio of rams to ewes is 1:40, with the rams running with the ewes for six to eight weeks to cover three oestrus cycles of 17 days duration. Lambs are marked at around four weeks of age. Marking involves ear-marking or branding to show who owns them, putting rubber rings on their tails and testicles to cause them to fall off, or else cutting them off. Lambs are weaned at three to five months.[1]

The main aim of genetic improvement in Merinos is to increase fleece weight while maintaining or reducing fibre diameter. Other aims include reducing skin wrinkle because of susceptibility to fleece diseases and fly-strike, and selecting away from sheep with a lot of face cover because this trait correlates with poor reproduction performance.

---

1. Mulesing is a traditional technique to help control fly strike of sheep. It involves removal of a strip of skin from each side of the lamb's tail. The skin here grows close and leaves the area of skin around the tail relatively bare and tight, thereby greatly reducing the possibility of blowfly strike around the crutch. Any apparent hardship caused by mulesing is markedly less than the agonies of fly strike. This practice has become controversial and subject to concerns about animal welfare. Mulesing is being phased out of the industry. New methods of controlling fly strike are being developed, including breeding animals with bare breeches.

The extent to which superior performance of genetically superior sheep is realised depends on environmental conditions. For instance, if feed is not plentiful, average types of sheep may perform as well as superior sheep. If feed is plentiful, the potential production superiority of the genetically superior animals will be realised. Increased production from genetic gain will not be realised unless the feed supply and other environmental factors are also improved.

Milk

Milk is composed of water, fat and non-fat solids. The non-fat solids are proteins (mainly in the form of casein), milk sugar in the form of lactose, and minerals. The objective of dairy farming is to produce as much milk of a particular quality as possible from the feed, the cows and other available resources. To do this, dairy cows are usually mated so as to produce a calf every year so that they produce milk for about ten months and are then dried off for about two months before calving.

Milk production by dairy cows follows a distinct pattern over the whole lactation period. Milk produced rises for about six weeks after calving and then gradually declines. This pattern is called the lactation curve. The height that this curve reaches at its peak, at about six weeks after calving, dictates the total production for the entire lactation, as the shape of the lactation curve for a well-fed and well-managed cow is relatively fixed. Feeding the animal just before and after calving is important in achieving high levels of milk production from the entire lactation. While voluntary food intake increases in the cow to provide nutrients for milk production, this is usually insufficient to meet the nutritional requirement for milk synthesis in early lactation. The shortfall in nutrients is met by significant mobilisation of body tissues, that is, the cow experiences a period of severe negative energy balance.

How dairy farmers produce milk

Dairying is carried out in every state of Australia, but is confined to the relatively small areas where conditions are suitable. Dairy production under pastoral conditions requires large quantities of high-quality forage available for a large proportion of the year. Dairying takes place in those areas of Victoria, Tasmania, NSW and South Australia that have high rainfall and long growing seasons, or that have access to irrigated pastures. Most forms of dairying obtain up to half the feed required from outside the farm, in the form of purchased grain, fodder and silage. The dominant breed is the high-milk-yielding Friesian–Holstein, which makes up over 60 per cent of all dairy cattle. About

15 per cent of the national herd are Jersey–Friesian–Holstein cross-bred cows, and about 25 per cent are Jerseys.

Depending on the prices received for milk at different times of the year, dairy farmers may produce milk all year round by calving at different times of the year. Alternatively, a seasonal pattern of milk production might be followed, in which all cows are dried off prior to calving in the most seasonally favourable time. Commonly in these situations cows are dried off in the winter when feed is short, to calve in spring when feed is plentiful. However, consumers want milk all year round so it is produced by some farmers all year round. Producing milk in the less favourable times, in terms of pasture supply and management, is more difficult and more costly. But, as supply is less at this time, a premium price is paid to producers for their 'out-of-season milk', thus making it worthwhile for some farmers to produce milk at the more difficult times of the year.

A well-managed cow with the right genetic make-up and getting enough high-quality feed, pasture and some concentrates, will produce 5000 litres of milk or more per year. This would equate to over 200 kg of butterfat per year, at 4 per cent butterfat content. To produce this quantity of milk over a year and to be pregnant again and have two months dry, a reasonable-quality cow will require the nutritional equivalent of about 5 tonnes of cereal grain or 10 tonnes of dry matter of good-quality pasture. With concentrate-feeding good cows will produce 6000 litres of milk per year, which at 4 per cent butterfat content produces 240 kilograms of butterfat per year.

The major potential health problem in dairying is mastitis, which is controllable by good herd husbandry. This involves antibiotic treatment of infection, dipping teats in antibacterial solution after milking, treating cows with antibiotics on drying off and getting cell counts of milk from individual cows. Bloat remains a potential health problem in Victoria and Tasmania where white clover makes up a significant component of dairy pastures. Acidosis is a problem in cows where supplementary feeding strategies are poorly managed, especially when feeding cereal grains in large amounts at one time, as in bail feeding, and as significant proportions of the total diet.

The farmer's main challenge is to provide the high quality and quantity of nutrition needed to produce large quantities of milk from the whole business. Most of the milk comes from pasture, with varying proportions of supplementary feed. Improved highly digestible and nutritious perennial pasture grasses and legumes in heavily fertilised rain-fed and irrigated pastures form the basis of milk production. Supplementary nutrition comes from concentrates such as cereal grain and stock meals, fed routinely or strategically, depending on feed supply, and from fodder crops and forages such as lucerne, maize or oats grown to meet peak demands and to fill seasonal gaps in feed supply.

Artificial breeding and recording to assess the production of cows and then the genetic merit of sires form the basis of much genetic improvement of dairy herds. Over half of all dairy cows are artificially inseminated and nearly half are herd recorded.

The average size of dairy herds increases steadily, and is at around 250 cows. The number and size of larger-sized dairy operations are increasing quite rapidly, with many milking 500 cows using rotary-design dairies, and usually using family and hired labour or operating on a share-farming basis. Dairy businesses running large dairy herds of 1000 or cows or more are increasing too.

Most commonly a self-replacing system is used to maintain animal numbers and improve genetic merit. Most cows are artificially inseminated (AI) with semen from good quality bulls, though bulls are used to mate any cows that either do not become pregnant using AI or are not worth using AI on. Heifers of good genetic quality (AI bred) are mated at 15 months to calve and enter the milking herd at two years. The alternative replacement system is to buy heifers in calf. Annual milk production of cows will increase each year for the first four or five lactations as udder glands grow with each lactation. Milk production then declines with age. Average-performing cows will be culled at six or seven years of age. Female calves are retained and grown to mating age. They are selected into the herd or sold as springing heifers. Male calves are usually sold at birth.

Intensive controlled-environment animal activities

Pigs

The profitability of an intensive animal activity such as pig production is determined basically by reproduction rates, conversion rate of feed into meat and rate of growth (of meat, not fat). There are two major breeds of pig: Large White and Landrace. These make up over 85 per cent of all pigs. Commonly Large White × Landrace crosses are used as first-cross dams in commercial herds. These are then crossed back to either a purebred sire such as a Large White or Landrace, or to a cross-breed sire, to reap the benefits of hybrid vigour.

Young reproductively mature female pigs, called *gilts*, are mated at about six months at the second oestrus cycle, weighing 90–100 kg liveweight. Seven to 14 days before mating, feed is increased by 25 per cent to increase ovulation rate and litter size. But feed needs to be reduced to pregnancy level after mating, otherwise embryo mortality rises. The duration of oestrus in the sow is one to four days, with an average of two to three days. Ovulation occurs on average 36 hours after standing heat. Optimum insemination time is 12 hours before ovulation (that is, 24–36 hours after the sow is first detected to be in oestrus). Reproductive performance factors, such as ovulation rate and litter size, are affected by the age of the sow, with both factors increasing as the sow age increases. Generally sows are culled after six litters, when they are around 3.5–4 years old.

Gestation of the sow is 114 days (three months, three weeks, three days) and average litter size is around 11 pigs weaned. The sow's milk is low in iron, and piglets are prone to suffer anaemia. Therefore, the piglet needs an injection of iron in the first few days of its life. At two to three weeks of age, the piglet starts eating dry feed as well as sow's milk. After parturition and during lactation the feed requirements of the sow increase markedly, as much as 0.5 kg per day per piglet. Generally weaning is at four weeks, sometimes six weeks. At this stage, the piglet weighs around 12 kg and the feed conversion ratio for the next eight to ten weeks is high until the pig is around 45–50 kg liveweight. Around 2.5 kg of feed puts on 1 kg of weight. The feed conversion ratio steadily declines to 4:1 as the pig approaches 90–100 kg liveweight at around 140 days of age. An average feed conversion ratio for an entire pig herd is around 3.25 kg of feed per kg liveweight of meat produced. Feed accounts for 70 per cent of the cost of producing pig meat, with labour costs another 10–15 per cent. In some piggeries, growing pigs are fed *ad lib* (as much as they want) up to 40–50 kg. Beyond this weight, feed is rationed in quantity, and a lower percentage protein feed is given, in order to achieve a lean carcass. Other piggeries feed *ad lib* until the pigs reach sale weight at around 90 kg liveweight by 140 days. Carcass dressed weight is around 70 per cent of liveweight. A set of realistic production targets for a commercial piggery are:

Pigs born per litter	12.5
Live pigs per litter	11
Percentage of pigs born dead	4.4 per cent
Pigs weaned per litter	11
Litters per sow per year	2.3
Pigs born per sow per year	28.75
Post-weaning mortality	3 per cent
Pigs sold per sow per year	22
Average liveweight at sale	90 kg
Herd feed conversion ratio (liveweight)	3.25

As a pig grows, the composition of its body changes, and the rates of growth of bone, muscle and fat changes. Following weaning, the rate of growth of muscle or meat is far greater than the rate of growth of the skeleton or the rate of growth of fat. As the animal gets bigger and heavier, the rate at which fat is put on increases. By the time the animal weighs about 110 kg liveweight

(80 kg carcass weight), the rate of gain of fat is about twice that of muscle. These relative rates vary with the breeds and strains and sex of pigs. However, pigs, like all biological systems, have an S-shaped growth curve. A pig that averages a growth rate of 500 g per day will have experienced growth rates per day greater than 500 g per day earlier in life and less than 500 g per day later in life.

Pigs require the following substances in their diet: energy (carbohydrate or fat), protein, calcium and phosphorus. Some amino acids (building blocks) can be synthesised by pigs from protein, whereas others cannot be. The amino acids that the pig cannot make are called essential amino acids and must be supplied in the diet. The major one of these is lysine, as it is the first limiting amino acid in cereal-based diets for pigs. If a deficiency of an amino acid exists, then the other amino acids cannot be fully utilised. Furthermore, surplus amino acids cannot be stored, so they need to be supplied daily in the diet. Pig diets are formulated on a ratio of digestible lysine:digestible energy. The major sources of protein or amino acids in pig diets are cereals, protein concentrates such as meat and fish meal, and free amino acids, of which lysine is the only practical one.

Pigs are raised intensively. Housing and environment are critical factors in production. The environment in which the pig lives has a major impact on the key measures of production performance: feed conversion, weight gain, reproductive performance, animal health and animal behaviour. Newborn piglets have little subcutaneous fat and little capacity to produce metabolic heat so are very susceptible to cold temperatures. Growers and breeders have little fat. Temperature has to be maintained near the optimum to avoid the pigs using excessive feed just to maintain body heat. Heating is needed. Growers and breeders need an environment where temperatures are constrained between 15 and 25°C.

Disease can break out at any time in intensive animal activities, with disastrous effects. Piglets are prone to enteric diseases caused by strains of *Escherichia coli* (*E. coli*), and by Rotaviruses and other respiratory diseases. Pigs are particularly susceptible to disease outbreaks because they are housed closely together. Where the cause of the disease is bacterial, antibiotics can be effective. A wide range of other diseases must be prevented or managed. Disease problems associated with any form of intensive animal production can be tackled by operating as a disease-free or minimum-disease activity.

Recently in Australia there has been some move from intensive to less intensive production systems. These can vary from sheltered, group-housed pigs on deep litter to outdoor production systems, which may partly address some of the issues relating to the capital cost of housing and restricting movement. However, there are still many issues related to poorer performance and disease that have not been resolved.

Good economic performance of a pig activity is a 90 per cent conception rate, four-week weaning age, ten piglets weaned per sow and the sow being pregnant again within seven days of weaning. This level of performance produces about 24 pigs per sow per year.

Poultry for eggs

The objective in commercial egg production is to get as many eggs as possible from the investment in the pullet and the feed over its laying life. The key to the highly productive modern egg production is that it is done under very controlled environmental conditions, using highly genetically improved egg layers. In most commercial egg-production operations, laying hens are kept in cages where they are watered and fed and eggs are collected. Manure is dropped out of the cage. From a production viewpoint, there is an optimum amount of space required for each bird, and if space per hen is reduced beyond this amount, reduced production and increased mortality follows.

Generally the replacement chicks needed to be grown to pullets to replace old hens are bought in. Specialist hatchery operations produce chicks for laying operations. Sometimes egg producers will buy chicks and rear them until they are about 20 weeks old and ready to go into the laying flock. Pullets start laying eggs at 20–22 weeks old. Egg-laying and body growth are competitive uses of nutrition.

Reproduction in poultry is sensitive to day length. Decreasing day length causes hens to stop producing eggs. Thus, light is an essential management tool in commercial egg production. Fourteen hours of daylight stimulates hens to start laying eggs. It is standard to keep egg-laying hens under a constant 14–16 hours of light from 22 weeks onwards.

Once the young hens start laying eggs, they will peak at 85–95 per cent production. That is, on any day during the peak production period, the number of eggs produced will be 85–95 per cent of the number of hens. Over its production life of about a year, a hen will lay about 270 eggs.

The aim of the egg producer is to convert a given amount of feed into as many eggs as possible. The nutrient and feed requirements of laying hens are well defined, and feed accounts for about 80 per cent of total egg production costs. A hen will require about 1.8 kg of feed to produce a dozen 50–60 g eggs. Poultry are very sensitive to their nutritional needs because their digestive processes, and respiration and circulation, are relatively rapid, in comparison with mammals. Also, poultry have body temperatures that are 5°C higher than other animals. This is why poultry grow rapidly. Rapid growth makes their nutrition critical.

Poultry require water, carbohydrates, fats, proteins, trace elements and vitamins. The major requirement of a poultry diet is energy. This comes from cereal grains. Poultry need animal or vegetable protein as a source of the amino acids that are needed. As well, a poultry diet based on energy from cereals and animal or vegetable proteins needs to have added calcium, phosphorus, sodium chloride, manganese and iodine, plus vitamins plus some non-nutritive feed additives to promote health and growth.

Under natural conditions, at the end of the breeding season, hens shed and renew their feathers, called moulting, and cease laying eggs. After moulting, they will start laying again. Under intensive production conditions, the first moult usually occurs at about 16–17 months after hatching, when the hen has been laying for 10 or11 months. It is usual to replace hens at this age, after laying for about a year, because young hens lay at a higher rate than do older hens.

One way to get hens to lay more eggs over a given time is to force them to moult. This is done once egg production has declined to an unsatisfactory level. Moulting causes the hens to stop producing eggs. Hens can be forced to moult by reducing the light they receive to eight hours per day. Feed is then removed for ten days, following which the normal production feeding and lighting regime are reinstated. Hens will return to 50 per cent of production in six to eight weeks. They will then lay at a higher rate than they would have for another few months. The objective is to get more eggs from the hen over a given time than she would produce if the first production and moulting cycle was left to run its course without interference and forced moulting. Whether forced moulting is used or not depends on the gain from the extra egg production and the value of the hen at the end of her 'forced' second production cycle compared with the return from a lower total number of eggs and the value of the hen after one production cycle.

Successful poultry production is based on the control of disease. The behaviour of birds is also an important factor influencing production and influencing decisions about such things as housing design and space provided per bird. Birds have a well-defined social order of dominance called 'pecking order'. This seems to be a mechanism for keeping the peace, as birds submit to rank instead of resisting, and also as a means of 'keeping the space', as birds defend the space they need in front of their face.

The practice of intensive housing of hens has resulted in growing community concern about animal rights and animal welfare. The essential concern is that the hen is given a happy life, or at least a happier life than they appear to have when crowded into a cage to do nothing else but produce an egg every day. It is difficult to be sure about what makes a hen happy, but it is noteworthy that hens that are not in a satisfactory physiological state will not reproduce.

Poultry for meat (broilers)

The modern advances in poultry genetics and nutrition have combined to make it possible for birds, bred specifically for the purpose, to convert feedstuff into poultry meat, very efficiently. The broiler industry comprises hatchery egg producers who provide the commercial broiler chicks that broiler growers then feed until the broilers reach sale weight. Broiler chicks are highly heterozygous, being cross-bred to achieve much hybrid vigour. Broilers grow very rapidly and reach a weight of around 2 kg dressed by six weeks. Feed conversion rates of around two units of feed to make one unit of meat are achieved. About 4 kg of feed will produce a 2 kg broiler. Feed costs account for about 80 per cent of total costs of production.

REVIEW QUESTIONS

1. Explain how a farm animal can be thought of as a cyclical biological system within a cyclical farming system.
2. How do farmers attempt to match animal feed supply and demand through the year?
3. Why might you say that operating an intensive animal industry is like sitting on top of a volcano?
4. Compare and contrast the key elements of a prime lamb versus wool production system. What are some of the key management decisions that a farmer needs to make for each system?
5. Discuss some of the key management challenges facing dairy farmers trying to improve milk yields and the efficiency of milk production.
6. It could be argued that nutritional management of animals in an intensive production system is more challenging than nutritional management of animals in a grazing situation. Discuss.
7. Explain why the goal of producing beef or lamb with particular carcass specifications does not always go hand-in-hand with available resources.
8. Why are production targets important for intensive production systems?

FURTHER READING

Australian Agriculture: The Complete Reference on Rural Industry, National Farmers Federation, Canberra, 1993.

Cottle, D. J. (ed.), *Australian Sheep and Wool Handbook*, Inkata Press, Melbourne, 1991.

Feeding Standards for Australian Livestock: Pigs, CSIRO Publishing / PISC (SCARM), 1987.

Feeding Standards for Australian Livestock: Poultry, CSIRO Publishing / PISC (SCARM), 1987.

Freer, M. and Dove, H., *Sheep Nutrition*, CSIRO Publishing/CABI, 2002.

Gardner, J., Dunkin, A. and Lloyd, L., *Pig Production in Australia* (2nd edn), Butterworths, Sydney, 1991.

Nutrient Requirements of Domesticated Ruminants, CSIRO Publishing, 2007.

Reid, R., *A Manual of Australian Agriculture* (4th edn), Butterworths, Sydney, 1990.

Sainsbury, D., *Poultry Health and Management* (3rd edn), Blackwell, London, 1992.

Squires, V., *Livestock Management in the Arid Zone*, Inkata Press, Melbourne,: 1981.

WEBSITES

Australasian Farm Business Management Network: www.csu.edu.au/faculty/science/saws/afbmnetwork/

CSIRO: www.csiro.au/

National Farmers Federation: www.nff.org.au/

NSW Department of Primary Industry: www.dpi.nsw.gov.au/

Vic Department of Primary Industry: www.dpi.vic.gov.au/dpi/index.htm

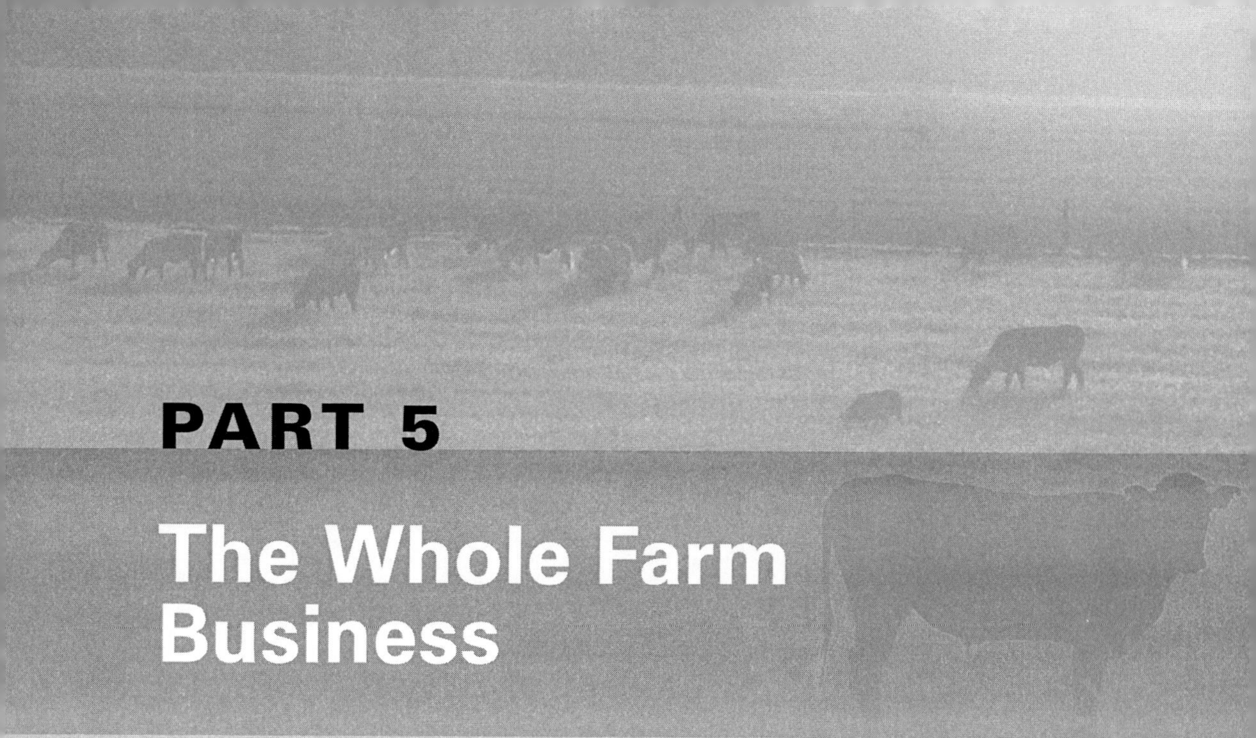

PART 5

The Whole Farm Business

Whole Farm Analysis: Principles

8

CONTENTS

Introduction

A farm business is made up of a mix of people, natural resources, technology, economics, and finance. Farmers buy their inputs to production and sell their output in markets. What they do is subject to laws and affected by things governments do. Running a farm business means dealing with lots of risk. There is risk about how much crop and livestock they will grow and what quality they will be. There is risk about what production will cost and how much the production will be worth. There is risk affecting how much cash will be available to pay interest to people who have lent funds to the farm business. There is risk about things governments from all over the world might do that will affect how farm businesses are able to operate.

To understand farm businesses well, four main principles of a technical, economic and financial nature, and one method or approach to analysis, must be understood. The four principles fundamental to understanding how farms work and make profit and grow are:

- the *principle of comparative advantage* in production
- the *principle of diminishing marginal returns* to extra inputs to production
- the *principle of increasing financial risk* to capital invested in the farm
- the *probability principle*, which affects all decisions and their outcomes.

The one approach to understanding how and why farm systems operate the way they do is the *whole farm approach*.

The four fundamental principles and the one approach to farm management analysis—the whole farm approach—are the keys to understanding farming. These are explained below.

Four principles and the one approach to farm management analysis

Principle of comparative advantage

The principle of comparative advantage plays the key role in determining whether or not a particular form of production is profitable in an economy. This principle holds that farmers in particular regions or countries are best off when they do what they are relatively best at doing, and then trading.

The principle of comparative advantage has existed as long as people have used resources to produce things they need. Consider the following simple example of some very early, very simple production systems.

Suppose with three months work primitive Farmer A could rear an animal ready to eat. With four months work this farmer could produce a tonne of grain to eat. One animal and one tonne of grain will feed herself for a year. She works for seven months and has five months in which to do other things, or to do nothing at all, which she regards as a good thing too.

A neighbouring primitive Farmer B living in the next valley has better conditions for agriculture. He can produce an animal ready to eat with two months of work, and can produce a tonne of grain with one month of work. To produce the meat and grain he needs for a year, he only has to work for three months, leaving nine months free for him to do other things, or nothing at all, which he also values highly (see table 8.1 below).

TABLE 8.1 Comparative advantage: two farmers

	ONE ANIMAL	ONE TONNE OF GRAIN
Farmer A	3 months' work	4 months' work
Farmer B	2 months' work	1 month's work

It takes Farmer B two months of work and Farmer A three months of work to produce an animal. Farmer B is 1.5 times better than Farmer A at producing one animal. It takes Farmer B one month of work to produce a tonne of grain. Farmer A takes four months of work to produce a tonne of grain. Farmer B is four times better than Farmer A at producing grain. Another way to think about this situation is that in the same period of time, Farmer B could make 1.5 times more animals and four times more grain than Farmer A could do.

Really useful economic ideas are not always obvious. In this case, the surprise is that even though Farmer B is better than Farmer A at producing both meat and grain, Farmer B is still better off getting his meat from Farmer A. It is all because of the principle of comparative advantage. Farmer B could work for two months and produce two tonnes of grain. Farmer B could then swap a tonne of grain with Farmer A in exchange for one animal. Specialising in producing grain, and trading with Farmer A, means that Farmer B has only to work for two months in the year to obtain the total annual supply of animals and grain he needs to feed himself. He would have to work for three months if instead of specialising and trading he grew all the grain and all the animals he needs.

Farmer A could work for six months to produce two animals; one of the animals she would exchange for a tonne of grain from Farmer B. When Farmer A specialises in growing animals and trading with Farmer B, Farmer A feeds herself fully for the year. The benefit is that to obtain all the food she needs in the year, she only has to work for six months instead of seven months.

By Farmer B specialising in grain production and trading with Farmer A, and Farmer A specialising in animal production and trading with Farmer B, both farmers are able to feed themselves for the year more cheaply (efficiently) than if they each produced all of their own grain and animals. This is the key result of the application of the principle of comparative advantage.

In the example, specialisation in activities and trading has made both producers better off. The principle of comparative advantage makes the case against trying to produce everything you need yourself—called being self-sufficient—and makes the case for specialising in doing some things really well and trading with others who do different things really well. The key to the gains from applying the principle of comparative advantage to decisions about production is that by doing what you are relatively best at doing, and then trading, all parties are made better off than otherwise would be the case. Specialising in what each business is relatively best at, and trading, is the basis of economic activity to improve the lot of the growing population of the world.

A consequence of the operation of the principle of comparative advantage is that what farmers do and the way land is used is determined not by what a producer is best at doing on their farm, compared with what their competitors are best at doing on their farm, but by what a farmer is *relatively* best at doing on their farm compared with what their competitors and their trading partners are *relatively* best (or least bad) at doing on their farm.

For example, farms in a higher-rainfall area (Wet) could be better than those in a drier area (Dry) at both livestock and cropping (or recreational farming) (see table 8.2).

TABLE 8.2 Livestock and cropping in Wet area compared with Dry area

WET	DRY
Crop in Wet compared with crop in Dry	
Crop in Wet is: better than Dry by a little	Crop in Dry is: worse than Wet by a little
Livestock in Wet compared with livestock in Dry	
Livestock in Wet is: better than Dry by a lot	Livestock in Dry is: worse than Wet by a lot

In the situation in table 8.2, crop and livestock activities can be done better in the Wet area than in the Dry Area. This means the Wet area has an absolute advantage in both crops and livestock over the Dry area. However, this does not mean that farmers in the Wet area produce all the crops and livestock and farmers in the Dry area produce nothing at all.

The operation of the principle of comparative advantage would dictate the situation in table 8.3.

TABLE 8.3 Livestock versus crop for Wet and Dry areas

LIVESTOCK VERSUS CROP IN EACH AREA	
Wet	Dry
Crop is better than livestock by a little	Crop is better than livestock by a lot

Dry is much better at crop than livestock, by a big margin. If Dry was to give up growing crop to grow livestock instead, it would give up a lot. Wet is only a little better at crop than livestock. If Wet was to give up crop and grow livestock instead, it would give up only a little.

This means, in terms of the production systems of the Dry and Wet regions, farmers in Dry have a *comparative advantage* in crop compared with farmers in Wet. This is because farmers in Dry are a lot better at crop than they are at livestock compared with the situation for farmers in Wet, who are only a little bit better at crop than livestock. Farmers in Dry are *relatively* better at crop than livestock, compared with farmers in Wet.

So farmers in Dry specialise in crop and farmers in Wet specialise in livestock. The net result is that more crop and livestock are produced in total from the two regions than would be the case if farmers in Wet and Dry produced both crop and livestock on their farms. Farmers in both Wet and Dry regions are better off by specialising in what they are *relatively best* at doing.

This idea—specialising in what you are *relatively best* at and trading with someone else who is relatively worse at something else, and everyone gaining—applies to all activities in an economy. It is the basis for international trade. It is why Australia is a large exporter of its agricultural products, and a large importer of manufactured secondary goods. The outcome is that overseas buyers get cheaper and more farm products, while Australians get cheaper and more manufactured secondary goods.

Principle of diminishing marginal returns

Let us return to the example of Farmer A and Farmer B outlined earlier and take a closer look at the way Farmer B grows grain. He has worked out that applying animal manure to his crop as fertiliser (probably obtained from his trading partner, Farmer A) enables the crop to produce more grain. Using trial and error over time, Farmer B establishes that different amounts of fertiliser added to the crop are associated with different amounts of grain produced.

Interestingly, he finds out that at first each extra unit of fertiliser adds a greater amount to total output. Then as more units of fertiliser are added, the extra output from each extra input diminishes, even though it still adds to total output. Farmer B eventually finds out that if he adds one more unit of fertiliser the total amount of grain declines. He has used too much fertiliser and made the crop sick. Less fertiliser (input) would make more grain (output). The information Farmer B has about fertiliser input and grain output is called a production response function.

Farmer B has discovered the second big principle about the economics of production: the principle of **diminishing marginal returns**. This is a very good thing to know. If he did not understand this idea he could have thought that all he had to do to increase production of grain would be to keep on adding fertiliser and output would continue to increase. He may have thought that he could feed the whole world with grain from his little plot of land!

Suppose that after time passes, primitive Farmer B becomes a commercial farmer. He buys his fertiliser in a market and sells his grain in a market. He quickly realises, now that the inputs have a cost and the extra output brings in income, that it means the right amount of fertiliser to use is not simply the amount that makes the maximum total grain that is possible. No. The fertiliser costs $1 per unit and should only be used if it adds more than $1 of grain. With diminishing returns to extra units of inputs of fertiliser, eventually there comes a level of fertiliser use where he spends $1 on another unit of fertiliser and this adds only $0.50c worth of grain. He loses $0.50c on this unit of input, and total income from grain minus total cost of fertiliser is reduced by $0.50c.

The maximum difference between total income from grain and total fertiliser cost (profit) is when the last unit of fertiliser costing $1 adds $1.01c worth of grain. Up to this point each unit of fertiliser has added more income from extra grain than it cost. Farmer B has discovered that the amount of fertiliser to use that gives the maximum total income minus total cost (maximum profit) is the amount of fertiliser that adds one cent more income than the fertiliser costs.

The description above is of the application of the fundamental principle of production, called the principle of diminishing marginal returns. This principle means that as extra inputs are added to production, such as fertiliser added to a crop, the extra production that is created becomes less with each extra input. In agriculture, inputs to production are land, labour, fuel, fertiliser, chemicals to control weeds and pests, medicines for animals, feed for animals and so on.

The operation of the principle of diminishing returns is so critical that processes of agricultural production and the behaviour of farmers cannot be understood properly unless this principle is understood properly. Consider another example. The initial units of an input that can be varied and added to a fixed input, such as fertiliser (varied) added to land (fixed), can initially have the effects of increasing quantities of output with each input. This is called the *increasing returns stage* of production.

Then, as more input is added, the extra production that results diminishes with each extra unit of input. Extra production is achieved, however, and total production continues to rise. This is called the *stage of diminishing extra (marginal) returns*. As even more inputs are added, each extra unit of input causes total production to decline. This is called the *stage of negative marginal returns*. More inputs cause less total production than is achieved using less inputs.

Simply knowing that production will respond to inputs in the way outlined above tells farmers a lot about how much input to add to their production. Obviously it makes no sense to use more inputs to make less production, even if the input is free. The stage of negative returns is the 'too much' level of inputs.

The stage of input use where extra inputs add increasing amounts of output per input, the increasing returns stage of production, is the 'too little' level of input. Farmers should use inputs at least up to the level where the average production that results from an extra unit of input has reached a maximum and is starting to decline. In between the 'too little' level of input and the 'too much' level of input is the 'just right' level of input. This rule holds true as long as the product is worth producing at all. That is, regardless of the costs of input and the value of the output, as long as it is worth producing, then the inputs used should be in the range defined as the 'just right', or diminishing marginal returns, level of input use.

FIGURE 8.1 Principle of diminishing marginal returns

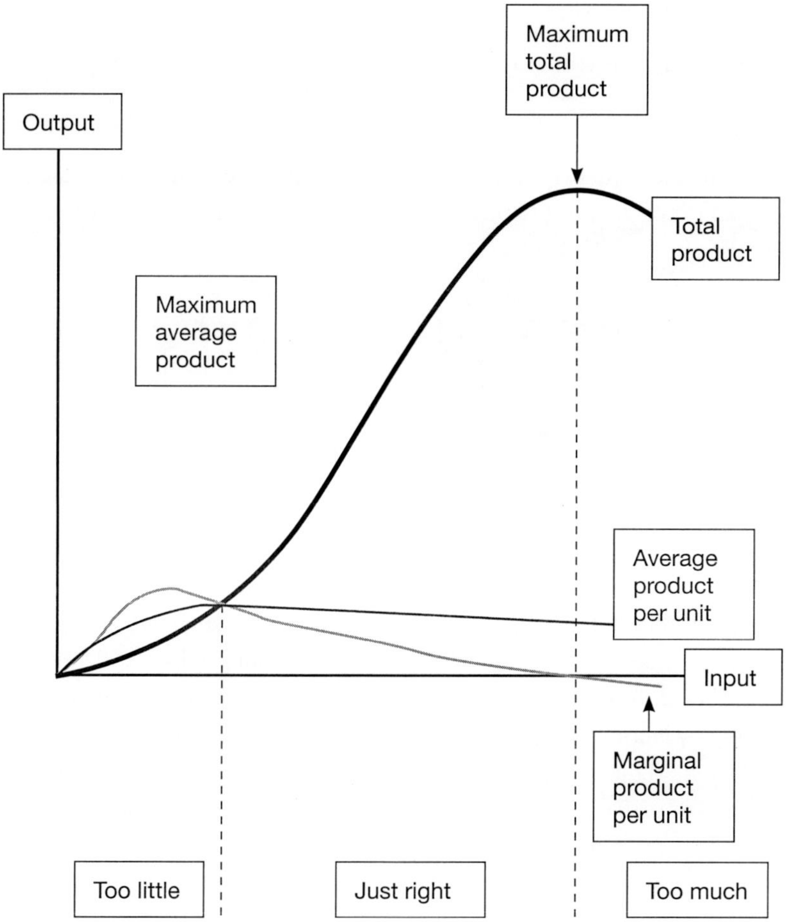

The precise best level of inputs to use in agricultural production is the level where the value of the last extra unit of production brings in just a little more than the cost of the last unit of input, e.g. the last unit of fertiliser used costs $1 and adds extra output that adds $1.01 income. This last unit of input adds $0.01 to profit. All the previous units of input added more than this amount to profit. The maximum profit is earned when the last unit of input just makes a profit. This idea sometimes confuses, causing people to think how is it that profit can be maximised when only a tiny profit is made on the last input. The point is that by using other inputs previously and up to this level, the producer has earned bigger profits on all the earlier inputs. Adding all the profits from each input gives total profit from all inputs. Another way to think about it is if one further input is used, costing $1, and this input produces an extra $0.50c income, the producer loses $0.50 from using that last input. Total profit has been reduced by $0.50. Production is in the 'too much' zone of input use.

The principle of diminishing marginal returns dictates that farmers wanting to maximise profit should use variable inputs:

- at least to the level where average production from each unit of input is at a maximum
- not beyond the level where **marginal** production becomes negative
- at the level within the 'just right' zone of production where the extra returns added from an extra unit of output just exceeds the extra cost of adding the extra unit of input.

The principle of increasing financial risk

The capital available to a farm business from profits is limited. Therefore, farm businesses have to borrow to expand. The principle of increasing financial risk applied in agriculture is that as a farm business uses more borrowed capital the chance of losing its own capital increases. The operation of this phenomenon is shown in the following example.

> A farm business has total capital of $10 million, made up of $5 million debt and $5 million equity. Operating profit is 10 per cent p.a. on total capital, an operating profit of $1 million. Interest costs 8 per cent p.a. on $5 million, so $400,000 interest is paid. Net profit is $600,000. If there is no tax or consumption uses of net profit, net worth or equity increases by $1 million – $400,000 = $600,000, a growth rate of $600,000/$5 million = 12 per cent. Now consider the alternative situation where the business makes a 10 per cent p.a. operating loss, a loss of –$1 million. The business still has to meet its interest obligation on the debt of $400,000. The total loss is $1,400,000. This represents a decline in equity of $1.4 million/$5 million, a decline of 28 per cent. The rate of growth of equity when things went well (+12 per cent) is not symmetrical to the rate of decline of equity when things went badly (–28 per cent), hence the principle of increasing risk.

The principle of increasing risk is about the obvious phenomenon whereby as the debt of a business increases relative to equity, the more risk there is of that business not being able to meet its debt servicing obligations at all times. It is also about the phenomenon that if businesses have a certain amount of debt and use the debt capital such that the return on total capital exceeds the cost of the debt, then the owner's equity grows at a faster rate than would have been the case if the business had a lesser amount of debt. The principle of increasing risk explains why businesses cannot simply keep borrowing and growing. The reason is that as the proportion of debt in total capital increases, so too does the risk. So as debt increases, capacity for further borrowing decreases and, at the same time, cost of further borrowings increases because it is more at risk. The business faces a trade-off between borrowing and equity growing rapidly when things go well (earnings exceed cost of debt) and putting equity at risk of declining when earnings are less than cost of debt, and with equity declining at an even greater rate than it grows when things went well. Hence, increasing debt means increasing risk. The further complication is that if a business does not use debt, and use it well, the business will not grow sufficiently rapidly to remain competitive and equity will be eventually lost because of this, too.

As the size of a farm business increases and the uncertainty and potential losses increase, the cost of borrowing too increases. As debt increases, farmers find they cannot borrow all that they want at current or possible interest rates because their business is a riskier proposition for lenders. The operation of the principle of increasing financial risk limits the size of farm businesses.

The farmer has to strike a balance between expected profit and amount of debt and interest. This balance will vary for each farmer at any time, and will change for any farmer at different stages of their farming career. The effects of uncertainty and the principle of increasing risk on the attitude to risk of investors means that it makes sense for a farmer not to borrow and farm 'to the limit'.

The probability principle in farming

The probability principle in farming concerns farmers not being able to know for certain what is going to happen when they grow and sell a crop or some animals or animal products. Almost everything farmers do has a range of uncertain, possible outcomes. The answer to the question 'if I do this and this and this, what will be the result?' is 'it depends'. The ultimate result of a planned production process depends a great deal on some important but unknowable factors:

- how much will it rain, and when?
- how much will yield be, and of what quality?
- what problems will there be—diseases, timeliness of operations, quality of operations, accidents, labour issues?
- what prices will I receive for what proportions of production?

Farmers form judgments about various levels of prices, yields and costs that are most likely to occur. If they think that the overall result has a high enough chance of happening, considering how they feel about the consequences if it does not happen, then they will go ahead. They know, however, that they are always and unavoidably taking a gamble, making a bet with every decision. This is the essence of farming. The core of the farm management task is trying to make good bets that produce results that the farmer wants to achieve.

When a change is made on a farm, the final outcome will not be precisely what was thought most likely to happen at the time the change was made.

Numerous different outcomes are possible. Analysing changes and making decisions requires taking account of the volatility of potential outcomes. There are ways of doing this formally. First, personal probabilities and distributions of levels of key variables can be used in **budgets**. Second, budgets of 'what if' scenarios (sketches or models) can be built using a computer spreadsheet. The results are contemplated in terms of how likely they are to occur. Less formally, decision makers use their intuition ('gut-feeling' or judgment). All of these approaches are useful when evaluating the merit of making a change on a farm.

Approaches to decision making that are more formally structured than acting on intuition and lessons from past experience, or acting randomly, can give a better chance of making the right decision. The structured approach involves making sure that all relevant data is considered and a consistent and logical process is used to select the action or set of actions most likely to reach the goal(s). To do this, goals must first be defined clearly.

The most striking feature of the best farm managers is their mastery of information—this seems to be the key to success. Managing risky businesses is about gathering relevant information, weighing it judiciously and acting accordingly. It must be poor information that causes some farmers to take the common but misguided view that when things are good, it can only get better, and when things are bad, it can only get worse.

Understanding probabilities is useful when making decisions on proposals that are risky. Probabilities are strengths of belief about an event happening, and are expressed as a rating from zero (no chance) to one (certainty). Probabilities can be stated as the odds of an event happening. An event might be thought to have a probability of occurring of 0.4. This means that it is thought that there are 4 chances in 10 of it happening, and 6 chances in 10 of it not happening. In terms of odds, this is a 6/4 against chance. If it was thought to have 0.6 chance of happening and 0.4 chance of not happening, then this would be a 6/4 on chance, that is, it is less than 1/1, an even money, 50:50 bet. Something regarded as fairly certain, having probability of happening of 0.8, would have 8 chances of happening, 2 chances of not happening. This is a 4/1 on chance, about as good a bet as you could get. Odds of 100/1 against means that the person who is offering the odds believes that there is one chance of the event happening and 100 chances of it not happening.

To convert odds to probabilities the formula is:

$$\frac{1}{1 + \text{odds}}$$

Thus, with odds of 10/1 = 1 chance in 11 = 1 ÷ (1/11) = 0.09 probability

The 100/1 event has a probability of occurring of one in 101, or 1/101, which is 0.0099. The 9/1 against chance has a probability of happening of one in 10, or 1/10, which is 0.1. The 5/4 against chance is thought to have 5 chances in 9 of not occurring, and 4 chances in 9 of happening, a probability of occurring of 4/9, which is 0.44.The basic rule is that the probabilities assigned to two or more related events must add up to 1.0 (100 per cent). When a coin is thrown into the air, the probability of it falling heads upwards is one out of two, that is, 50 per cent or 0.5. The likelihood of it falling tails upwards is also one out of two, or 0.5.

It is usual to expect an event with a 50:50 chance of occurring to happen half of the time. That is, if someone had a fair coin and were to toss it a lot, it could be expected that half heads and half tails would show. Suppose that after a number of tosses there have been more heads than tails. Then someone might back tails in the hope that the next toss would show a tail. They would be mixing up two things. First, the undeniable truth that ultimately 50 per cent heads and 50 per cent tails will happen. Second, the reality that as the percentage of each outcome approaches 50 per cent, the absolute difference between the number of heads and of tails can still be very large. This is because a large number of tosses would have had to have been held. The bettor can still end up losing, even after a large number of tosses of the coin. In terms of the total number of losing bets the bettor could still have backed many more losers than winners. The percentage of each outcome that occurs will certainly be edging towards 50 per cent, but the divergence from 50 per cent of each outcome (say 49.5 per cent heads and 50.5 per cent tails after a large number of events), will be a small fraction of a large number. It can all add up to more losing bets than winning ones, and could still be a large number of losing bets. Hence, the number of losing bets could be still very large indeed. The 'gamblers fallacy' occurs because the 'law of averages' or 'the correction factor' works only over a very, very large number of events, so much so that the probability for each single toss of the coin remains the same. Many of the events about which farmers have to assess probabilities are like this. For practical purposes, the probability of a good season following a bad season is roughly the same as the chance of a good season following a good season, or a bad season following a bad season. The coin does not remember.

A technique for analysing risky decisions is to calculate expected values of possible outcomes. For example, if there are 1000 tickets in a lottery and the prize is worth $1000, then each ticket has an expected value of $1000 multiplied by 0.001 chance of winning. The expected value of a ticket is $1. The expected value of buying all the tickets is $1000. Expected value is worked out as the sum of a series of possible outcomes such as 'good', 'most likely', 'poor' seasons or prices. These are then multiplied by the probability of them happening, as estimated by the decision maker. So instead of using a single value for yield of, say, 2 tonnes per hectare, probabilities can be used to work out an expected value for yield. For example, suppose the farmer considers there is 0.20 chance of a crop yielding 0.5 t/ha, 0.3 chance of yielding 1 t/ha, 0.25 chance of yielding 1.5 t/ha and 0.25 chance of yielding 2 t/ha. The expected yield is:

$$
\begin{aligned}
(0.2 \times 0.5\text{t}) + (0.3 \times 1\text{t}) + (0.25 \times 1.5\text{t}) + (0.25 \times 2\text{t}) &= 0.1\text{t} + 0.3\text{t} + 0.375\text{t} + 0.5\text{t} \\
&= 1.275\text{t}
\end{aligned}
$$

If the most likely figure were used it would be 1 tonne per hectare as this is expected to occur 3 years in 10. Expected value is more useful than historical average because historical average relates to a unique past, never to be repeated exactly, while expected value relates to what is believed might happen in the new, different and unknown future. The expected value is a composite figure that captures the situation that would apply if all the chances came at once. Or, it is what you would get, on average, if you made the same investment a large number of times, and the probability distribution of outcomes remained stable. If this was to be true then profit over time will be greatest if the alternatives having the highest expected money values (EMVs) are always chosen. There are a number of limits to the usefulness of probabilities and EMVs. These are associated with the relatively small number of the main 'events' (seasons, prices, yields, investment decisions) involved in a farmer's career, when the 'laws' of probability work best over a large number of 'events', and the consequences of some events are far more serious for the success or failure of the business than other events. The small sample of 'events' means that good or bad 'luck' in terms of important variables in farming life have a significant influence over success or failure.

In farming much cannot be known. This is called uncertainty. Probabilities cannot be attached to uncertain events. While probabilities can be used to give a more realistic measure than single value estimates, when using probabilities 'the probability of the probability' problem arises. That is, the assumed distribution of probabilities is treated as though it is itself a certainty. More problems arise because the assumed relationships between the uncertain variables of an activity are also uncertain. There are some computer techniques that can be used to simulate runs of different seasons and prices to estimate probability distributions of outcomes on farm profit, finances and growth.

The probability principle emphasises that the future cannot be known and many outcomes are possible. People interested in agriculture have to think in terms of, and live by, the concept of distributions of possible events occurring, such as rainfall, yields, prices, costs and incomes. Many things could happen but only one thing will happen. Living by the probability principle means eschewing the use of historical averages when thinking about and making decisions. An average of a range of outcomes such as yield, price, cost and income is merely the sum of all the past levels of these outcomes divided by the number of them that we have added up. This has no particular meaning for any particular production year. In any year any of the range of outcomes, or combinations of them, could happen. People studying agriculture, or practising it, need to consider more and less likely events in any single production year, not an average single figure that can be achieved only over a run of many time periods. This is the probability principle.

The one approach: the whole farm approach to understanding agricultural systems

There are two broad approaches to thinking about questions in agriculture. One approach is to think in detail about a component part of the systemic activity of producing agricultural products that can help identify questions and problems to do with that part. This is the disciplinary specialisation that is necessary for progress in science. Thinking about the whole system and its component parts and drawing on a mix of disciplinary knowledge is another approach. Understanding agriculture requires both approaches. The whole farm breadth of knowledge about the system is required to identify questions and problems and put them in their proper perspective. Disciplinary and inter-disciplinary depth of knowledge is required to correctly answer questions and solve problems. Partial approaches, such as looking at a farm system and seeing only the soils or plants or animals or people, or emphasising the quantity of production (physical output) that results from the mix of resources (inputs of land, labour, capital) used, are not sufficient on their own.

The central theme of this book is that understanding production on farms means understanding both the parts and the whole of farm systems. Emphasising the whole farm approach means recognising that the goal of farm systems is to help meet farmers' goals. Achieving high levels of technical efficiency, resulting in high yields, is fundamental to farmers achieving their goals, but maximising technical efficiency is not 'the goal'. This is because maximising yield (technical efficiency) in a production period is not the same as making the most profit, or achieving more of a mix of goals in a single, or any, time period.

A whole farm business can be broken into parts: people, technical and physical elements; economic and finance aspects; risks; and phenomena beyond the farm gate such as markets for inputs and outputs, government policy and the non-farming public. The technical factors determine what is possible on the farm. Economics indicates what is most profitable and finance indicates what cash is involved. The human element is about what the farmers want out of their farming; whether the farmer can carry it out; and how much of the rewards to take in dollars. Forces beyond the farm gate help determine ultimate outcomes.

The starting point to the whole farm approach to understanding agriculture is recognising that every farm business is unique. First, the people running them are unique, with their own set of hopes and goals, skills, views of risk, stage of life, and family situation. Second, the physical resources of a farm business are different for every farm. Third, potential income and costs, and debts of each farm business—the overall financial situation—are different for each farm. Fourth, each farm has a unique history. The history of a farm business has determined the present situation of the business, and partly determines opportunities and directions for the future.

Emphasising 'the whole' in agricultural production means starting with the people, applying farm-level science and economics, and also at times going beyond the farm gate. The systems approach will sometimes encompass the whole from producer to consumer, or from individual farmer to the community and nation to which they belong. At other times boundaries need to be put around the sphere of interest: the crop system, the animal system, the feed system and so on. It is necessary to do so to avoid the trap of being unable to say anything about anything unless you can say everything about everything.

REVIEW QUESTIONS

1 What are the four main principles that determine how a farm business operates and whether or not a farm business succeeds? Briefly outline how these principles work.
2 Maximum yield (technical efficiency) is not the same as maximum profit (economic efficiency). Explain.
3 What does the term 'whole farm approach' mean? How do you apply the whole farm approach in practice?

FURTHER READING

Blackburn, A. and Ashby, R., *Financing your Farm: A Practical Guide to Financial Growth* (4th edn), Australian Bankers Association, 2006.

Boehlje, M. and Eidman, V., *Farm Management*, John Wiley and Sons, New York, 1984.

Malcolm, B., Makeham, J. and Wright, V., *The Farming Game: Agricultural Management and Marketing*, Cambridge University Press, Cambridge, 2005.

WEBSITES

Australasian Agribusiness Review and Australasian Agribusiness Perspectives: www.agrifood.info/

Australasian Farm Business Management Network: www.csu.edu.au/faculty/science/saws/afbmnetwork/

Australian Agricultural and Resource Economics Society: www.aares.info/

Australian Bureau of Agricultural and Resource Economics: www.abareconomics.com/

Australian Farm Institute: www.farminstitute.org.au/

National Farmers Federation: www.nff.org.au/

NSW Department of Primary Industry: www.dpi.nsw.gov.au/

Vic Department of Primary Industry: www.dpi.vic.gov.au/dpi/index.htm

9 Whole Farm Analysis: Practice

CONTENTS

Introduction

The criterion by which the performance of a farm system is judged is simple: to what extent does it meet the goals of the owner and any other interested parties, such as the public, who may be affected by the operation of the system.

People running farms are trying to achieve a number of goals that, taken together and allowing for inevitable trade-offs between goals, and over time mostly achieved to some degree, will amount to them having a good life. For example, a farm family may have goals of producing enough output to sell and earn income that exceeds the costs of the inputs, leaving a profit. Making enough profits in the short term will help the family to do important things like live and educate children and improve the quality of their resources. Over time, profits will increase wealth, which in turn will make it possible to achieve other objectives. Related objectives include survival and growth of the business in the face of competition from other farmers in Australia and around the world.

Owners of farm systems have many goals, not all are equally important and not all are able to be achieved. Some goals contradict others. There have to be trade-offs in the extent to which various competing goals are to be achieved.

For example, staying in business by paying all the bills will be one goal of a farmer. This is called *liquidity*. Another goal could be to have some very good, very long holidays. It is likely that these two goals will be in competition. The farmer will not be able to do both of these things to the full extent he or she would really like to do them. A compromise will be needed. Another goal may be to create sufficient wealth to enable a comfortable life in retirement and to give offspring a good start in life. Another goal may be to increase the size of the business in order to stay in business and be able to keep competing with competing farmers, which is called **growth**. Another goal could be to farm as well as can be, which is called *efficiency* and is measured as return on the total capital invested. All of these goals have an economic dimension. Economic goals can be summarised as:

- growth in wealth (change in equity)
- liquidity (net cash flow)
- efficiency (return on total capital).

Along with these goals are others such as having a business with a level of risk that is not too upsetting and being able to have some time for non-farm activities and leisure. Reducing risk and increasing leisure may have to be traded off against growth, cash and profit.

A focus on size of a farm business, however measured, is not the appropriate criterion to use when judging the performance of farm systems. From an economic perspective the relevant criterion is how efficiently the resources employed in the farm business are used, when the cost of using those resources in farming is valued at the **opportunity cost** (that is, what those resources could earn in their alternative uses). Judged this way the efficiency or contribution of resources used in farming can be compared with what those resources could be contributing if they were used in some other way.

The whole farm approach involves looking at all aspects of the farm system in detail. Applying this approach involves establishing the total resources that are being used in the farm system and determining their value. Each activity then needs to be analysed in detail to determine its performance. The sum of the performances of all activities is then used to obtain whole farm performance in light of the resources used.

The measures required are:

- total assets at start and end of year
- total debts at start and end of year
- total equity at start and end of year
- total profit from a year of operation
- total cash in minus cash out over the year of operation.

A helpful way for farm managers to start improving the performance of their farm is to work out how the business might perform in the coming year if it is run in one or more alternative ways. It can help first to establish how the business performed in the year just past.

When the situation of the whole farm has been established for the year just past and for the coming planning period, as it currently operates, the next step is to consider changes that may improve the performance of the farm. The techniques to use to consider the merit of changes are:

- compare the profit of the new whole farm plan with the profit expected from the current farm plan, and check if the extra profit is a good return on the extra capital that has to be invested to make the change, or
- use a **partial budget** where the extra benefits and extra costs of the changes in the new farm plan are evaluated. If the extra benefits exceed the extra costs of the change by an amount that gives a return on extra capital invested that is attractive, then the new plan should be adopted.

Profit and cash flow is estimated for the way the farm is currently being operated, or as it is likely to perform over the next few years. Using the whole farm approach and estimating changes that might potentially increase profitability and cash flows requires that a number of techniques are mastered. These include the techniques of constructing activity budgets, whole farm budgets, partial budgets and risk analysis of changes to the system. It is necessary to be able to estimate profit; and to be able to assess the state of the business' assets, debt and growth in wealth. It is also necessary to know how to estimate future cash flows in and out of the business and how to take account of risk in financial management and decision analyses.

Farm performance

There are two key questions of interest to farm managers:

1. How 'healthy' is the business in the light of future possible developments?
2. What are the best changes that can be made to maintain and improve the 'health' of the business?

There are two major challenges facing today's farmer:

1. how new technology can be profitably incorporated into the existing business organisation
2. how to be sufficiently flexible, mentally and financially, to adjust resource management to meet both changed economic circumstances and widely varying climatic conditions.

The whole farm approach to appraising the performance of a farm business is explained in the following sections.

Human considerations

Farm management analysis and decision making is about identifying correctly the real nature of the problem. The critical skill is identifying the true nature of problems and bringing to bear on them technical, human, economic and financial conceptual and analytical skills. Pertinent information, processed by sound analytical and planning techniques, makes for good decision making. Making decisions means evaluating and acting on information. The most outstanding feature of the best farm managers is their mastery of information concerning the whole spectrum of the process called farm management.

Goals

Two questions worth addressing are: what are people's needs and wants, and of these, to which can the concepts of farm management economics be usefully applied? In any consideration of the management of a farm the starting point is with the needs, wants and goals of the owner-operator and farm family.

The objectives of farmers loom large in any consideration of why farmers do or do not do something. The goals of the farm family are major determinants of how a property is managed and how the farmer might exploit the potential of the farm. Common goals of farmers are to overcome challenges and to farm well, and be recognised for so doing; to improve the physical state and appearance of the farm; to acquire extra land or control a bigger business for the future and for heirs; to have a reasonable but not profligate standard of living; to earn enough profit to be able to improve and develop the farm so as not to have to work so hard as they age; to achieve capital gain and increase wealth; to have good-quality animals in good condition; to reduce income tax; to have a satisfying rural way of life; and to have their children well educated. Importantly, goals are often in conflict; thus trade-offs are involved between the degree to which various goals are achieved.

A comprehensive listing of the goals and objectives of people owning and managing farms would include the following aims and hopes:

1. to ensure that the business survives and grows
2. to have a reasonable standard of living compared with others in society
3. to increase wealth over time
4. to improve the farm in its appearance and its long-term potential for productivity
5. to set up offspring in farming

6 to have an enjoyable rural way of life
7 to be recognised as a good farmer
8 to provide good-quality products
9 to avoid losses of income and capital
10 to not have too many worries from risk
11 to accept a challenge and conquer it
12 to achieve financial independence and security
13 to have enough leisure time
14 to reduce taxes paid
15 to educate family and themselves
16 to engage in community service
17 to be a respected member of the community
18 to be sufficiently financially secure to enable the pursuit of important non-farming interests.

When the goals of farmers are being considered in the light of the potential of the farm, the aim is to determine whether the product markets and the resources of land, technology, capital, credit and skills are compatible with these goals. Identifying where the manager's strengths, weaknesses and preferences lie in performing the various physical and mental tasks helps explain the way a farm business is being operated and how well it is performing.

The skills of the manager(s) and labour force are one of the main inputs to farm production. The quality of the human element is a major determinant of the potential profitability of the business. If a program of intensifying or developing the potential of a farm business is being contemplated, the farmer and the labour force need to have the physical skills and mental attributes to handle the changed situation, which is usually more technically complex and often more risky than the present circumstances.

Farm management

Deciding how to run a farm involves a series of steps. Problems, or needs for action, have to be recognised; the nature of the problem or need, and its effects on other aspects of the business, have to be defined accurately; and courses of action have to be chosen that are feasible in terms of the decision makers' objectives, the technical requirements and constraints, and the dictates of the markets for inputs and outputs. Analysis of the economic and financial feasibility has to be done. The non-measurable aspects of the decision have to be weighed up. Then a decision can be made. Once the decision is made, the intended steps have to be carried out well. As the results never turn out exactly as expected, the operation of the farm business then has to be monitored, controlled and revised in the light of the inevitable unexpected developments.

The main management functions of a farm manager are: (1) planning and deciding; (2) organising resources and doing the things that have been decided; and (3) keeping a close watch on how things are going and responding to the changes that arise continually. If nothing changed there would be no need for management. Planning would be sufficient because without change, planned outcomes would be the same as actual outcomes. As things change continually, management is about adapting to changes in circumstances, often before the changes have occurred, and when the full extent of the changes that are happening is not known.

The science and art of farm management has been the subject of considerable study over the past 50 years. Management cannot be assessed solely in terms of goals achieved. In part, management is about making the best decision in the light of information that is available at the time. Even so, a decision may not turn out to be the correct decision. Management decisions are made before the event. Whether it is the right or wrong decision can only be known after the event. A good decision can still turn out to be the wrong decision because of the way the many unknowns and uncertainties eventuate. Further, once a decision is made and action taken, the operation of the farm itself is different to that which prevailed before the decision. A new decision-making environment then exists for management.

The following are some characteristics of good managers:

- They know all the relevant information pertaining to whatever particular tasks or projects are currently at hand.
- They put a lot of emphasis on getting jobs done on time.
- They think ahead: a bit of paperwork can be a considerable aid here.
- They discuss ideas, procedures and alternatives with others.
- They make provision to cope with the unexpected.
- They have a system for keeping up to date with daily work achieved, for regular consultation, and for making sure that the employees have regular, clear instructions.

Part of farm business management is about the farm manager making good decisions. Decision making usually involves choosing between alternatives in the light of the decision maker's goals. The decision-making process has some generally recognised steps:

- recognising problems and the need for action
- making observations, collecting facts and getting ideas
- analysing observations and testing alternative solutions to problems
- making a decision
- acting on the decision

- implementing the decision
- taking responsibility for the decision
- controlling the implementation
- reviewing the outcomes and adapting the intended and expected to the actual
- doing it better next time, that is, learning from one's mistakes.

The importance of the human factor in business success cannot be overstated. Ultimately, it is the human qualities of individuals involved in conducting the business that will determine the results. For instance, having effective communication between the farmer/manager and employee(s) is fundamental since an absence of a proper mutual understanding makes the best of plans difficult to achieve. Having good communications within the family workforce is also important, but the means of bringing it about where it does not exist is usually difficult because of complex and close interpersonal relations.

Labour management is an essential part of farm management, and it is fruitful to look at such questions as:

- what motivates employees to work well?
- what are the secrets of the art of good labour management?
- how can parent–offspring problems that affect the operation of the farm be reduced?
- are employees justly rewarded for their labour?

Often the most vital aspect of management is the task of getting the right things done through people. Communication is at the core of the management of labour. When a farmer employs labour, the quality and timeliness of the field operations can be greatly reduced if the employee(s) lack motivation. Often, the main cause of lack of interest is that there has been no effective two-way communication of ideas and aspirations between the manager and the staff: the farmer sees things from one point of view, the employee(s) from another. The two parties may have different 'frames of reference' because their goals and view of 'reality' differ. Even though the perspectives, motivations and rewards of owners and workers differ, there are certain points of common interest, and it is around these that effective two-way communication can occur.

Technical conditions

Technical principles of using resources for agricultural production

Agricultural production is the process of using resources to make goods or provide services. Producers can use combinations of the three broad classes of factors of production—land, labour and capital—to produce one or many products. A key element in a farmer's decision about what to produce and how to do it is the objective of getting more, or the most, out of the limited amount of resources which he or she has to work with.

There are three basic relationships in production:

1 the relationship between the amount of a resource used and the amount of production (input–output)
2 the different ways resources can combine and substitute for one another in the production process (input–input)
3 the relationship between different products that can be produced with the resources that are available (output–output).

Feed supply—present

The two main sources of feed supply are pasture and the concentrate feeds used in intensive industries.

Concentrate feed

The chief issues in relation to concentrate feed are whether it is best for the operator to mix the rations himself or herself or to buy pre-mixed feed from a miller, how much labour is saved from using pre-mixed feed, and the relative quality of the two sources of feed. There is an increasing tendency to use pre-mixed feeds, and to use the labour saved to do other things such as run more stock.

Pasture supply

In an analysis of a farm's pasture supply, the following features need to be looked at.

Topography

The feed supply will vary with the slope and aspect of the land. An almost vertical slope facing south will obviously not have the pasture production of a river flat. Even so, a balanced combination of flats and hills can often provide more flexibility if grazing pastures in different seasons.

Soil types

Soils vary greatly in their structure and nutrient status, and thus in their fertility. They can be classified into a number of major and minor groups, according to their ability to produce pastures and crops. A breakdown of the area and range of soil types on a given farm is essential for physical analysis: it is the basis of both present and potential land use.

Present soil nutrient status and fertiliser history

Phosphorus and nitrogen are the commonest deficiencies in Australian soils. Soil tests can indicate the pH (degree of acidity or alkalinity), the levels of the major elements phosphorus, potassium and sulfur, and can be relied on fairly confidently for use in a fertiliser program. Soil tests for the significant trace (or minor) elements such as molybdenum, copper and zinc are not a reliable basis for predicting a response. Tissue tests of green growing plants are more useful. The fertiliser history is useful mainly as an aid to detecting deficiencies other than phosphorus. For example, if a farm has a history of having a total of 2.5 tonnes/ha of superphosphate applied over the previous 15 years, and the pastures are not vigorous, it is likely that some element other than phosphorus is lacking and so limiting production. In practice, an effective way of determining present fertiliser needs is a combination of soil tests every few years; records of past fertiliser use; some simple, moderately sized, trial plots; and the farmer's observations of how the paddocks have been responding to recent applications of fertiliser.

Pastures

The following aspects of the pastures have to be defined:

- native, introduced legumes, grasses, weeds
- balance between legumes and grasses
- density of plant population
- vigour
- suitability for various classes of livestock e.g. non-breeders, young growing animals, breeders, sheep or cattle.

Expected seasonal pasture production

Based on past rainfall records and farmers' experience, there is, for each region, an average expected pattern of seasonal pasture production, due to the rainfall pattern. Thus, in southern Australia, rain-fed pastures usually begin active growth in autumn, slow down in winter, reach a peak in spring, and do not grow in summer. The further north the region, the greater the tendency for pasture production to be summer dominant and winter dormant. Rainfall in Australia ranges from 20 to 3000 mm per year. Generally the lower the annual rainfall the more unreliable and variable it is.

The average seasonal pasture production has to be defined because this determines the system of animal production that is most profitably suited to the pastures. For example, in Victoria it is more feasible (but not necessarily more profitable) to have ewes lambing and cattle calving in autumn than it is in northern NSW, where winter pasture production is low.

Expected pasture supply

Only rarely does the average season occur. For example, in southern Australia, the opening rains, on average, occur during April. Yet they can begin as early as February, and as late as June. Winters there can be very wet, where the paddocks become waterlogged, so impeding pasture growth and use; or very dry and warm, thus producing good winter pasture supply. Similar variations about the average season occur in other areas, often with wider fluctuations than in southern Australia.

The strategies and the particular tactics which the manager of an individual farm adopts to handle these variations about the average, vary greatly. The general aim is to reduce the harmful effects of adverse situations, and to exploit favourable circumstances. The particular measures adopted by different managers vary greatly in their degree of profitability (or loss saving). Thus it is essential, when analysing a given farm, to learn what measures the operator adopts to handle variability, e.g. supplementary feeding, stock trading, fodder conservation, reduction in body condition of stock. Skill at handling the fluctuations about the normal pasture supply is one of the main requirements for the successful running of pasture-based animal production enterprises.

Feed supply—potential

In this part of the farm analysis, attention is focused on what potential exists for increasing the quality and quantity of feed grown. There are a number of avenues, which are discussed briefly below.

Fertilising pastures to increase quantity and quality

Fertilising pastures to increase quantity and quality has been the single most important action available to farmers who manage pasture-based animal production systems. There is much technical information to help them to carry out the physical part of a development program if they decide that it is economical to do so.

Fodder cropping

Winter and summer fodder crops can often be a valuable means of increasing the supply of high-quality feed at times of the year when feed from pasture is low. The scope for exploiting the potential contribution of fodder crops has to be examined in any farm analysis.

Developing irrigation

There is an increasing interest in using supplementary irrigation from water harvested from run-off that would normally go to waste, on pasture-based farms. As well, there have been a number of engineering developments, which mean that supplementary irrigation schemes can be run more cheaply than by using traditional methods. Generally, because of their cost, feed growth from supplementary irrigation schemes needs to be fed to animals who can convert it into a high-value product such as milk, stud stock, lamb or veal. It can also be used as hay or pellets as part of the ration of intensive animal systems such as pigs and poultry.

Feed conversion activities—present

In this part of the farm analysis, the aim is to see what system the manager uses to convert feed into money. On most farms, this is done through animals, but there are some enterprises that convert it directly into a saleable product such as hay or pellets. The potential for these activities seems to be increasing, as animal industries become more intensive, and so these forms of conversion are worth looking at. Let's now consider the chief features of animals as feed conversion mechanisms. The key aspects of analysis of the farm's animals are:

- type of animal activities
- structure of the flock or herd, by age, sex and numbers
- system used to replace the herd or flock.

There are two systems of replacement:

1. breed own replacements, as in a self-replacing flock of Merino sheep or Angus cattle
2. buy in replacement female breeders, as in second-cross lamb production using Border Leicester–Merino ewes, steer trading and fattening, egg and broiler production, and fattening porkers to baconers.

System of feeding

This can be:

- grazing (set stocking, rotational or both)
- hay, silage, fodder crop, grain agistment
- totally or partially fed on bought-in concentrate feed.

Husbandry

Husbandry covers use of medications and chemicals, veterinary services and the method of harvesting the product (milk, wool, eggs). It also includes the supervision and management at mating and birth and the time and interval (frequency) of parturition (calving, lambing and farrowing).

Genetic improvement

The avenues here include artificial insemination (AI), embryo transfer, selection based on performance recording, use of progeny-tested sires, cross-breeding, and purchase of special hybrid animals, as in the case of the intensive industries.

Feed conversion activities—potential

In the analysis of the animal-based feed conversion system, the potential and the scope for improvement can be considered under the headings used in the previous section. For example, under animal activities, it is necessary to consider whether there is a role for new activities, and also if any of the existing activities could be expanded or modified. With structure there may be potential for increasing the ratio of female breeding stock to males, for selling animals at a younger age, and for mating animals at a younger age. The replacement system may perhaps be improved by changing partly from a self-replacing system to buying in hybrid animals of superior performance, or by mating cull females to sires of another breed, thus producing sale animals that command a high price. The opportunity to improve performance by exploiting new or different systems of feeding is well worth investigating, because it is in this area of feed management that some of the greatest gains in profitability can be achieved.

Husbandry practices on many farms can often be improved by using the latest developments in the rapidly changing technology of control of internal and external parasites, in shearing, milking, poultry, shed and piggery design and operation, and in the development of minimum-care animal husbandry practices. Finally, capitalising on the various techniques for improving the genetic worth of animals is not very expensive. A little time and effort spent on investigating which of the genetic technique(s) may have a useful role to play on the farm being analysed often pays off.

Cropping activities—present

Types of crops grown

In broad-acre areas, the crops that are grown need also to be classified into winter and summer crops. On orchards and forestry projects, the age of the trees, and the expected life of each stand or block has to be defined.

Maintaining soil fertility

Most progressive managers aim to have a cropping system that yields consistently well, without at the same time reducing the long-term fertility and erosion status of the soil. However, there are many farms where the arable soil has been plundered. So it is essential when examining the cropping activities to find out what (if any) measures are being taken to maintain soil fertility. Fertility is defined in terms of structure, freedom from weeds, nutrient status, absence of harmful micro-organisms, and erosion. The question of how best to maintain fertility is particularly acute on intensive vegetable farms, but it is also crucial to orchard farmers and growers of plantation crops and broad-acre grain crops. The commonest method of maintaining fertility on broad-acre farms is to have legume–crop rotations.

There is a trend in some areas, where crops are more profitable than stock, to have non-legume systems of maintaining fertility. The techniques used include applying nitrogen fertiliser, stubble retention to maintain structure and reduce erosion, herbicides to control weeds, and chemicals to control harmful micro-organisms. Also, minimum-cultivation techniques using herbicides are increasing in favour. Assessment of the scope for expanding one or other of these two basic systems of maintaining fertility in the future is a vital step in any farm appraisal. On intensive vegetable farms, where there is little crop residue retained in soils, gypsum and poultry manure may have to be used to maintain structure.

Irrigation system

The chief aspects of the present irrigation system that need to be looked at are the amount of water available, water quality, salinity problems, labour requirements, system of delivering water to the growing crop, and drainage.

Cropping activities—potential

New crops

The first issue worth investigating is whether new crops can be grown on the area.

Change in mixture of crops

It may be biologically more advantageous to change the mixture of crops. For example, increasing the proportion of summer to winter crops may lead to better weed and disease control. Alternatively, increasing the area sown to stone fruits relative to apples and pears may lead to a reduction in labour demand for the whole orchard operation.

Extending length of cropping phase in the rotation

Since cropping is usually more profitable than grazing it may be technically possible, through the use of additional fertiliser and herbicides, to extend the length of the cropping phase in a legume crop rotation and still maintain fertility.

Irrigation development

On many farms, there is considerable potential for applying new developments in engineering and irrigation technology for cropping to achieve one or more of the following outcomes: more efficient use of water, reduction in labour, higher yields and greater areas cropped. The payoff for successfully adopting new techniques of irrigation can be quite high.

Machinery

There are a number of important questions to be considered in analysing the present machinery services of a farm:

- How are the machine services provided: by contract, share farming, leasing, ownership or a combination of them all?
- Are the machinery services available to do each operation properly and on time?
- What is the age, current value and the anticipated timing and cost of replacement of the machinery services?
- Do the implements match the tractors.

When considering the future, it is necessary to know such things as:

- whether a change in the form of acquiring machinery services would be profitable
- what measures could be taken to improve either the standard or timeliness of operations
- if it would be profitable to replace labour with more or bigger machines
- given that inflation in machinery costs is likely to continue, how the money will be found to replace machinery when the time comes
- whether the power source and the implements can be better matched.

Economics and finance

Economists concern themselves with the problem of how the relatively limited amounts of land, labour and capital in the world are used to satisfy people's needs and wants. Because there is not enough land, labour and capital in the world to go around and to provide all the goods and services which people need and want at any time, then choices have to be made about what is to be produced, what resources are to be used in which way, and who is to get the resulting output.

Analyses of questions about choices, or problems, cannot be useful if the focus is only on, say, the technical parts of the question while other parts of the question, which may be significant, are ignored, such as economic forces or social conditions. The philosophy of economic analysis is that it is more useful to solve the whole of a problem in an approximate manner than to solve a small part of a problem extremely well while leaving out significant parts of the question at hand. Thus economics and economic analysis is an interdisciplinary activity: the technical basis of farm economic analysis has to be sound, the human element included, and the institutional and political forces acknowledged.

In economic analyses the focus on maximum physical yield and on averages, commonly found in technical approaches to answering questions about resource use, is not a useful criterion for making decisions about resource use and production. The way to evaluate the health of a business is described below.

Business health

The essential economic criteria used in evaluating the health of a business (or **business health**) and the merit of a change to the business operations of a farm are:

- the contribution of activities to farm total gross margin (TGM) and profit
- actual recent and expected future net return on the capital invested, after tax and risk is allowed for in the calculations
- actual recent and expected future net cash flows after servicing debts
- the likely break-even criteria when a change is made, i.e. the performance of the 'new' situation needed to be as good as the existing situation or as good as another option
- the expected effects on debt and equity of a change as shown in the current and expected balance sheet of the business.

Activity analysis

An activity is a particular type of production, for example autumn lambing first-cross ewes, or wheat in a wheat-grain–legume crop sequence, or a Hereford breeding herd producing vealers, and so on. The gross margin (GM) for an activity is the **gross income** generated by that activity minus the variable (direct) costs incurred in earning the income from the activity. **Overhead costs** (or fixed costs) do not come into the analysis of the GM of an activity. The GM of an activity is calculated to identify the contribution the activity makes to farm TGM, that is, the sum of gross margins (GMs) from each activity on the farm. Then, TGM is the sum that is available to pay the overhead costs and make an **operating profit**. In working out the whole farm profit it is useful to think in terms of the contribution that the various activities have on overall profit. This is useful as long as it is remembered that *the performance of any* *activity in a farm plan is affected by the other activities also in the farm plan.* That is, the contribution of one activity to farm profit is in part a function of the other activities also on the farm. Once the whole farm is broken down into separate activities, some of the 'whole' is lost, and so too is some valuable information. This is summed up in the aphorism 'the whole is greater than the sum of the parts'.

Farm TGM is the sum of GMs of all the particular activities run together on the farm in their own particular way. Activity GMs calculated on the basis of GM per unit of some resource used, such as land area or feed supply or labour used or capital invested, can be compared with a view to expanding or reducing activities. Estimating the contribution an activity makes to TGM can be a useful device as long as it is remembered that the contribution to TGM of the GM of any single activity would be different with different combinations and sizes of activities on that farm. Comparison of activity GMs for different farms makes little sense.

Animal activity gross margin

An animal activity GM is animal income minus **variable costs**. Income from an animal activity is made up of sales of animals produced during the year, sales of animal products, sales of by-products, the difference between the value of old animals sold and the costs of replacing them, plus any change in value of the flock or herd due to a change in numbers from unsold production, or to a change in values of animals.

The variable costs of any animal activity are:

1 *feed*: maintenance costs of improved pastures, cost of forage, crops, hay, straw, silage, purchased feed, home-grown grains, agistment, direct or casual labour costs
2 *husbandry*: medicines, cleaning materials for milking shed, contract plus casual labour, and services (veterinary, shearing)
3 *marketing*: brokers' and agents' fees, transport, processing and selling.

The GM of any animal activity can be calculated as follows.

1 Calculate sales of animals (animals produced for sale, not cast for age) and animal products.
2 Calculate the increase or decrease in value of stock on hand from the start to the end of the year. This includes the value of animals not on hand at the start of the year but on hand at the end of the year because they have been produced during the year. It can also include any change in the value of the animals on hand at the start of the year. The values chosen here can have a major effect on income, for example, if a current market value was

unusually high or low. For this reason it is better to use the same values per head for the various classes of the stock on hand at the start and at the end of the year. The difference in value between animals which have been culled each year because they are now too old (called cast-for-age animals) and purchases of replacement animals each year to maintain the size of the herd or flock, gives annual **depreciation** or appreciation of animals. This is either a cost (depreciation) or an income (appreciation).

3 Calculate the value of sales of animal by-products.
4 Add 1, 2, and 3 to obtain gross income.
5 Add together the cost of feed, the cost of husbandry, and the cost of marketing to obtain total variable costs.
6 Subtract 5 from 4 to obtain gross margin.

To calculate an animal activity GM it is first necessary to estimate what animals go where within the system over the year. This can be done first with a sketch of movements of animals in the flock or herd, and the numbers then put into a livestock trading schedule.

FIGURE 9.1 Flock/herd diagram

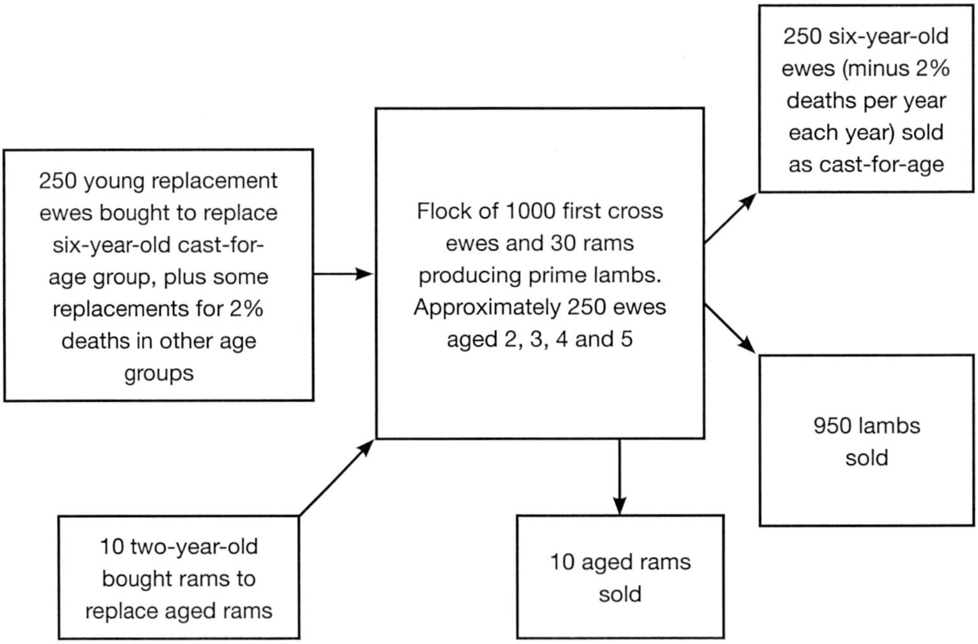

Livestock trading schedule

The livestock trading schedule is a record of the changes that occur in an animal activity through the course of a production period, and contains information about flock and herd trading profit or loss for the year. In a livestock trading schedule is recorded all the additions to and subtractions from the initial numbers and value of the animals in an activity for the period of interest (usually a year). A livestock trading schedule acts as a check on all of the movements into and out of an activity. It also acts as a record of changes in values of animals as they are bought, sold, born and as they age through the year. The livestock trading schedule then forms the basis for calculating trading profit from the animals in the activity. Trading profit from animals plus animal product sold makes up total income from the activity. A livestock trading schedule is shown in table 9.1. In it is recorded:

- animal type and class
- number and value at the start of the relevant time period
- births
- purchase numbers and value
- sales number and value
- deaths and rations
- transfers in and out of classes within an activity
- transfers in and out of activities
- numbers and value at the end of the time period.

Table 9.1 is an example calculation of livestock trading profit using a livestock trading schedule for a breeding flock based on bought-in replacements.

There is another way to calculate the GM of an animal activity that could be used if the total number of animals in the activity are in a stable state i.e. herd or flock are not increasing or decreasing over the year. Then the GM can be calculated using three components:

1. Gross income from sales of animals (whether animals produced that year or animals of a capital nature such as breeders) and animal products produced during the year.
2. Variable costs.
3. The difference between the value of animals cast for age and culled as part of the routine replacement system and the value of animals purchased as replacements. Animals purchased to replace animals that are culled are in reality a capital transaction, an investment—not a cost of production. Sales of capital items, such as old breeders, are not activity income. However, the overall effect of sales minus purchases of replacements is that for a steady state herd or flock the income or cost from capital depreciation or appreciation of the livestock capital of animals over time is captured by the value of sales minus purchases when numbers in the herd or flock is stable. The value of animals retained in self-replacing herds and flocks is automatically counted as a cost because the revenue which could have been had from selling them is forgone when they are kept and not sold.

TABLE 9.1 Livestock trading schedule: prime lamb activity

	NO.	$/HD	TOTAL VALUE		NO.	$/HD	TOTAL
OPENING NO.					Sales		
2 y.o. ewes	200	120	24,000	Cull ewes	196	30	5,880
3 y.o. ewes	200	120	24,000	Lambs	1000	80	80,000
4 y.o. ewes	200	120	24,000	Cull rams	10	30	300
5 y.o. ewes	200	120	24,000				
6 y.o. ewes	200	120	24,000				
Rams	30	400	12,000				
Births	1000						
				Deaths	20		
PURCHASES				**CLOSING NO.**			
Rising 2 y.o.	220	120	26,400	2 y.o.	220	120	26,400
Rams	10	800	8,000	3 y.o.	196	120	23,520
				4 y.o.	196	120	23,520
				5 y.o.	196	120	23,520
				6 y.o.	196	120	23,520
				Rams	30	400	12,000
Total	2,260		166,400 (A)		2,256		218,660 (B)

Trading Profit B – A = $52,260

An activity budget for a prime lamb activity is shown in table 9.2. A farmer has provided information as follows: flock of 1000 Border Leicester × Merino cross ewes are mated to prime lamb sires. Ewes lamb at two years old, are kept for four lambings, and are cast for age at nearly six years old. Lambs weaned and sold amount to a weaning rate of 100 per cent. Ewe deaths are 2 per cent in each age group. Replacement ewes are bought in at about one year old. No account is taken of changes in market values of stock not sold during the year because changes in market values can be transitory.

TABLE 9.2 Activity gross margin: prime lambs (NSW Department of Agriculture)

GROSS INCOME	$
WOOL	
1000 ewes @ 5 kg greasy @ $6/kg	30,000
30 rams @ 6 kg greasy @ $6/kg	1,080
Trading profit	52,260
Total income	83,340
MINUS VARIABLE COSTS	
Shearing and crutching	5,000
Animal health	5,000
Supplementary feed	10,000
Freight	3,000
Wool selling expenses	2,000
Stock selling expenses	6,000
Annual fertiliser	8,000
Pasture maintenance	3,000
Total variable costs	41,000
Activity gross margin	42,340
GM/ewe	42.34
GM/DSE	21.67

Crop activity gross margin

Expected crop GMs combine expected yield, price and variable costs for crop alternatives. Activity GM is the income from sales of output plus unsold or consumed stocks produced, minus variable costs for a given year. A GM for a crop comprises gross income (yield by farm gate price), assuming that output is sold in the year when the crop is produced, less cash variable costs. Activity GMs are usually expressed as GM per hectare. The variable costs of cropping include:

1 *growing costs*—these may be seeds, fertilisers, water, labour, sprays, and machinery running costs (fuel, oil, repairs)
2 *harvest costs*—such as casual labour, machinery running costs, harvesting materials

3 *marketing costs*—such as direct costs of storage, processing, transport, and selling costs
4 *machinery costs*—variable costs include repair and maintenance costs that cover costs from use-depreciation. (Depreciation cost due to time and obsolescence is an overhead cost, and occurs whether the machine is used or not.)

Table 9.3 is an example of a gross margin budget for a wheat activity.

TABLE 9.3 Wheat activity gross margin (per hectare)

GROSS INCOME	
3 t/ha @ $200/t	$600
MINUS VARIABLE COSTS	
Freight, levies etc.	$50
Seedbed preparation	$20
Fertiliser: Superphosphate Urea	 $50 $50
Herbicides and application	$90
Sowing	$10
Harvesting	$20
Insurance	$10
Total variable costs	$400
Gross margin per hectare	$200

Crop sequence gross margins

The expected GMs of individual phases of crop sequences are not sufficient information on which to base decisions. The way activities are analysed depends on why they are being analysed. The relative lengths of alternative crop sequences, and the size and timing of GMs of each phase of a sequence, have to be considered to validly evaluate alternatives. The effects of one crop in one year on another crop in another year have to be counted. These effects might be of benefit, such as providing a disease break or nitrogen. Or, it could be harmful, such as depleting nitrogen, or adding to yield-reducing or cost-increasing populations of disease-causing organisms, weeds or crop regrowth. In comparing the profitability of different farm plans involving different crop and livestock combinations, returns from entire sequences are compared, not individual segments of a sequence. If a long fallow is used to grow reasonable crops, the land has to be set aside for six months or more without producing anything except perhaps some short periods of grazing of unwanted grasses and other weeds. This means that the GM per hectare devoted to crop is halved, that is, one fallow hectare and one crop hectare are needed to produce each crop.

If it is assumed that each segment of each sequence were going to be present on the farm in each year, then the annual TGM per sequence-hectare is the figure to use to compare with alternative rotations. For example, a crop sequence might be one hectare of wheat followed by one hectare of peas. In any year total crop area is sown half to wheat and half to peas (see below). The profitability of a cropping sequence depends on the GM of the crop phases, the GM of non-crop fallow or pasture phases, and the relative lengths of the various phases of the sequence.

$$\begin{aligned}\text{Wheat GM/ha} &= \$180\\ \text{Peas GM/ha} &= \$220\\ \text{Annual GM} \div \text{Sequence ha} &= (\$180 + \$220) \div 2\\ &= \$200/\text{ha}\end{aligned}$$

Using activity budgets in planning a change

If an activity promises a GM of $180 per hectare and an alternative activity has an expected GM of $200 per hectare, it is not necessarily correct to infer that the most profitable step to take would be to implement the second activity in place of the first activity. Perhaps some simple changes could raise the expected GM of the first activity to $220 per hectare, whereas the alternative activity may not have any scope for such improvement. Or perhaps the first activity has complementary effects on another activity that makes it more valuable. It is necessary to first see whether the existing GM of an activity could be improved by the use of better technology or management.

To use GMs for planning (**gross margin planning**) the mix of activities on the farm, first select the activity with the highest GM per unit of the assumed most limiting resource (often but not always land). Next, expand this activity to the maximum level permitted by whatever limits more of it being done, such as land, labour, capital, rotation needs, preference for a mixture of activities to spread risks and so on. Other activities are introduced in order of decreasing GM per hectare. This continues until further increases in TGM cannot be achieved without exceeding the limits imposed by the resources available for this activity and any other limiting considerations. Note that activities often cannot be considered independently. For example, the complementary use of land with cash crops (such as wheat and sheep) needs to be considered, as does the feed provided by cereal stubble for a sheep activity. If complementary effects are not considered, the GM of the cereal activity is understated, and the GM of the sheep activity is overstated.

Activity gross margins are not the correct technique to use if a change in farm plan involves a change in overhead costs, as is most often the case. In this situation a partial budget of the change is needed.

Costs of machinery

There are a number of types of costs involved in any machinery operation. The first category of costs of owning a machine are called the **fixed costs** or overhead costs and include depreciation, insurance, registration, shedding and interest on the capital invested. The capital invested in the machine has an opportunity cost: the capital could earn interest in another use. All of these costs are incurred regardless of whether or not the machine is used.

The second category of machinery costs is called variable costs, which are direct costs of operating the machinery and include fuel, labour, tyres, lubrication, batteries, repairs and maintenance. These costs are constant per hour of operation and consequently vary directly with the hours of operation. Repairs and maintenance expenses per hour generally increase as the machine ages. A proportion of repairs done on machines is usually the result of the age of the machines as well as their use.

The third category of machinery costs are called penalty or timeliness costs. Operations not done at the best times because of the size or reliability of machinery can incur a cost to the machine's owner. This cost comes from weather losses at harvest, poorer yields, or a poorer quality of product than would have been achieved with better timeliness of the key machinery operations.

Depreciation

Most capital items depreciate in value. This is either due to wearing out or to obsolescence. A depreciation cost has to be counted to get the true cost of the business operations in a year. The simplest way to estimate annual cost of using capital items over a year is the straight-line method. This method has the assumption that an item loses value by the same amount each year.

Consider the case where a farmer needs to estimate the depreciation of some machinery in order to be able to estimate his annual profit. Suppose the machine is expected to last five more years and it is expected it would cost $100,000 to replace at the present time. After five years it is expected that the old machine will be able to be sold for $40,000 in current dollar terms. The annual depreciation costs are calculated as follows.

$$\begin{aligned}\text{Annual depreciation} &= (\text{Current replacement cost} - \text{Expected salvage value}\\&\quad\ \text{at end of life in current dollars}) \div \text{years}\\&= (\$100{,}000 - \$40{,}000) \div 5\\&= \$12{,}000\end{aligned}$$

Interest

Suppose a farmer was considering investing in a new machine and was worried about having a lot of money 'tied up' in it and thereby incurring an interest cost. The interest cost is estimated for the average value of the capital that will be tied up in the new machine, over the life of the machine. The initial value of the machine, plus salvage value divided by 2, gives the average value of the capital tied up. In this case if the machine is to cost $100,000 and be worth $20,000 after ten years in today's dollars, the average value of the capital invested is:

(Current value + Salvage value in current dollars) ÷ 2 = $60,000

A real interest opportunity cost of 5 per cent is charged

Interest = $60,000 × 0.05
= $3,000

Adding the fixed costs of registration, insurance, and shelter gives the total annual capital, interest and other overhead costs of owning a machine. Note: the numbers to use in the calculation of depreciation and interest depend on the reason for doing the calculation. The choice of numbers to use also depends on whether or not it is necessary to allow for inflation in the future.

To compare costs between alternative machines, all today's dollars and real interest rates can be used. To estimate how much to allow to replace machines at the end of their life, future inflated dollars and interest rates need to be used. To estimate annual profit, expected replacement and salvage values, current dollars can be used, and actual interest payments are relevant.

Annual use

The annual use or the annual output of a machine will determine the relative unit cost of a large or small machine. For example, if a larger tractor has an annual overhead cost of $8000 while that of a smaller one is $4000, and if the small tractor works 400 hectares of cultivation, its overhead cost per hectare is $10. If the larger tractor cultivates 800 hectares the overhead cost per hectare is also $10. If the larger tractor completes 1000 hectares of cultivation, the overhead cost per hectare will drop to $8 per hectare.

The less a machine is used, the higher the fixed costs per unit (per hectare or per hour). As the machine is used more, the fixed costs are spread over more hours or hectares, so the unit costs are lower. It will not be economical to own expensive machinery that will have little annual use. The amount of use determines whether it is cheaper to own the machine or hire its services.

Whole farm financial management

Owners of farm businesses are interested in three main things:

1 profit that is earned by the capital they have invested and borrowed: this measures efficiency, expressed as percentage return on total capital, operating profit/total capital × 100/1
2 cash that is available each year to pay all the bills, including paying interest on loans and repaying borrowed capital when it is due: the cash available is called liquidity or net cash flow
3 increasing their wealth, called growth: this is the addition to the capital the owner of the business has after all debts have been repaid.

The important measures of the financial performance of a farm concern efficiency (profit), liquidity (**net cash flow**), and growth (change in equity over time). This information is in the balance sheet and the profit and cash budgets. The relationships between the key elements of the financial state of a farm are shown in figure 9.2.

Balance sheet

The balance sheet records the total value of the assets and liabilities of the business at any time. Assets are things of value that the manager controls. Total assets are the total capital value of the things used in the business. Liabilities are debts or claims on the assets of the business. Equity is the value of the owner's share of the total capital invested in the business, that is, assets minus liabilities.

The balance sheet records all the claims on the assets of the business, which come from whoever provided the means used to acquire these assets. Equity in a business is the amount the owner can claim from the business. It is the amount the business owes its owners. However, claims on debt are stronger than claims on equity, because debts have to be paid in full before equity can be returned to the owners.

The information in the balance sheet reduces to a simple identity: total assets minus total liabilities equals equity (or net worth). The balance sheet gives a picture of the assets and debts of the business on a particular day of the year.

Assets in the balance sheet can usually be categorised as current, intermediate, and fixed. Current assets are expected to be converted to another form or used up within a year. They include cash, short-term bank deposits, accounts receivable and stocks of inventories such as grain, wool or trading livestock. Intermediate assets have a life of more than a year and up to ten

FIGURE 9.2 Farm profit, growth and net cash flow

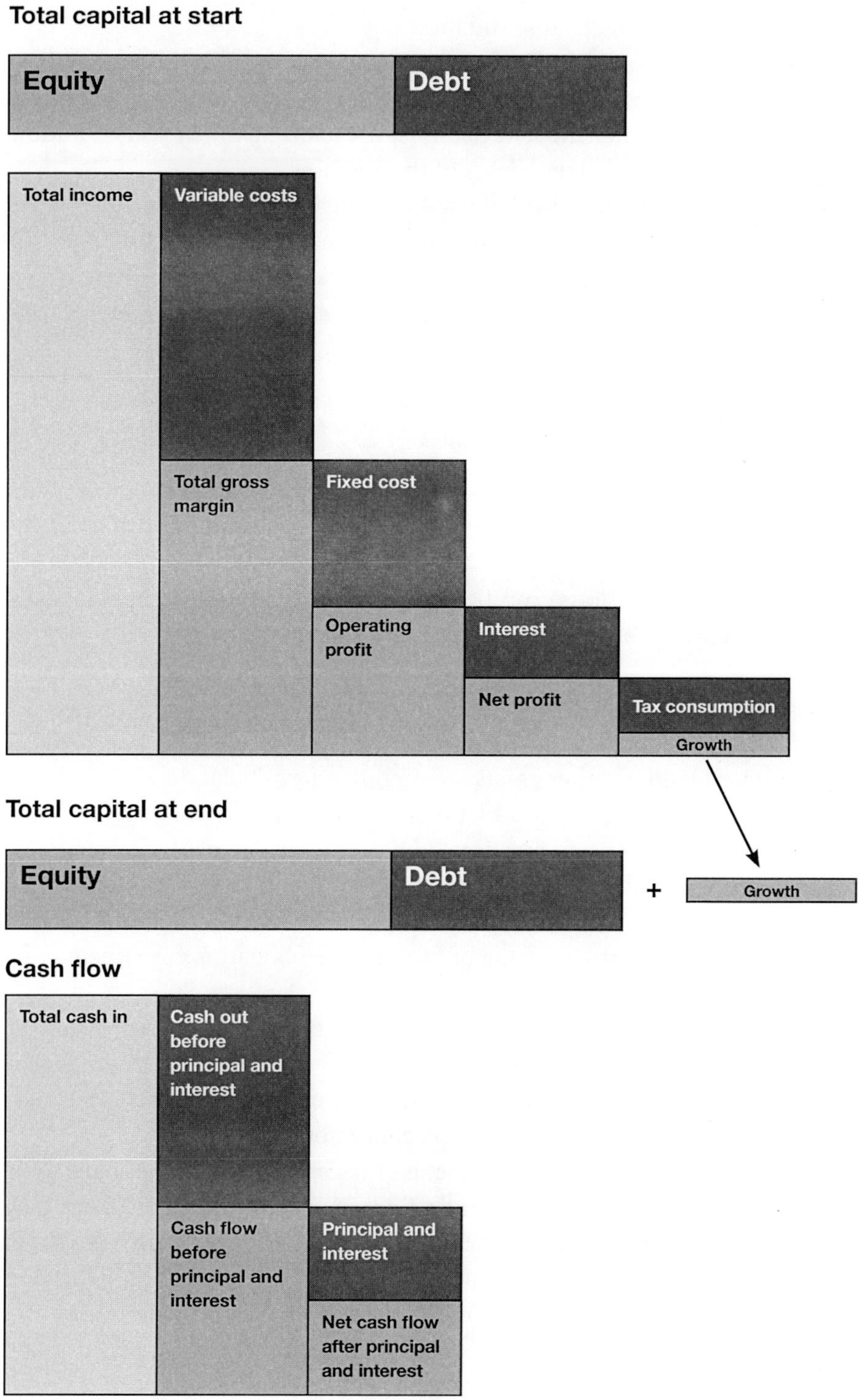

years, and include most plant and equipment, breeding livestock, or medium-term financial investments. Fixed assets (or **fixed capital**) have a relatively long life, such as land, improvements, and buildings.

The same categories are applied to liabilities. Current liabilities are obligations to others to be met fairly soon, within the year, such as a bank overdraft, and accounts payable. Intermediate liabilities are loans up to ten years duration. Fixed liabilities are longer-term debts of the business.

The value of the stock of capital used in a business varies over the period involved. The average value of all the capital invested in production over the period is used as an estimate of the amount of capital in the business over a year. The average value of capital is estimated as the opening value of total capital plus closing value of the capital, divided by 2, for example:

$$\text{Average value of total capital} = (\text{Opening value of capital} + \text{Closing value of capital}) \div 2$$

Profit does not indicate economic efficiency until it is related to the amount of capital used to produce it. Efficiency is operating profit expressed as the percentage **return to total capital** (operating profit/total capital). The rate of earning of the total capital can be compared to the rate of earning capital employed in some other income-producing activity.

$$\text{Percentage return on total capital} = \frac{\text{Operating profit} \times 100}{\text{Average value of total capital (WIWO)}}$$

When leased land or other assets are used in a farm business, as is increasingly the case, the value of leased land is included in the total capital being managed and operating profit is estimated before lease payments are deducted. Operating profit calculated in this way is the return to all the resources used and indicates efficiency of the whole business. If leased land is included in the balance sheet, the present value of future lease obligations are treated as a liability of the business.

Solvency and liquidity of the business are the other relevant aspects of the balance sheet. A business is **solvent** when assets are more than liabilities, that is, if total assets/total liabilities is greater than 1, the business is solvent. A business is insolvent if, were it to be sold, all debts could not be met.

Liquidity refers to the ability to meet all the cash demands that have to be met in the planning period. A test of liquidity is whether cash and near-cash (current bank deposits, government bonds and securities, saleable stocks of grain, wool, trading livestock) will be able to meet the interest on debts and debt repayments when they fall due in the short- to medium-term future.

It is informative to distinguish between changes in net worth and changes in asset and debt structure of the balance sheet. These changes have different implications for liquidity. For example, an equal increase and decrease in assets can come from a farm operator buying an asset such as machinery or livestock and paying for it with cash out of the bank. The structure of short-term assets relative to intermediate-term assets has been affected, and net worth is unchanged, but liquidity is changed. Or, taking out a medium- or long-term loan to pay off a number of short-term loans results in an equal increase and decrease in liabilities, leaving net worth unaffected. The debt has been restructured, but the annual debt-servicing requirements, and thus liquidity position, have been changed.

Profit

A budget of profit can be based on the returns and costs of a business for a past time period and show the profit that the business has made. Or, a budget of profit can be about the income and costs expected to occur in the future planning period. (Strictly speaking, a budget is about the future, whereas accounting statements are about the past.) Both types of budget of profit are useful. Major changes between the performance in the previous year or two, or changes between the past and the expected profit, can provide valuable lessons.

Profit budget

Gross income (this includes cash sales of produce plus non-cash changes in inventory value of livestock, stocks of output such as grain or wool)

minus Variable costs (usually these are cash costs of growing crops, livestock)

equals Total gross margin (this is available to cover overhead costs and make profit)

minus Overhead costs (this includes cash and non-cash costs, such as depreciation)

equals Operating profit (also called 'Earnings before interest and tax' (EBIT). This is the return to total capital).

The returns and costs in the profit budget have cash and non-cash items such as depreciation and inventory changes. The profit budget has all income earned from farm operations and all costs incurred in that period. Cash operating expenses include variable and fixed cash costs.

Expenditures for the purchase of capital assets are not considered an annual cash expense but investment, since these assets are used by the business for more than one year. Instead the costs of capital assets are allocated over their service life by including annual depreciation as the annual cost.

Interest payments on loans are not an operating expense. Financing arrangements are distinct from costs of the production activity. Interest payments are to do with financing the business, not production, and are a reward to those who lend capital to the business.

Estimates of profit include estimates of income and costs that are not actual cash receipts or payments. For example, such estimates might include the value of stock produced but not yet sold, or the amount of depreciation of a piece of equipment in a year. Most capital items depreciate in value over time, from wearing out through use or from obsolescence. An allowance for depreciation is deducted from gross income each year.

The reward for operator's labour and management is costed and deducted to estimate profit and the return to the capital of the business. The value of an owner-operator's labour and management is indicated by professional farm managers' salaries, or the next best alternative return an owner-operator could expect to earn elsewhere (called opportunity cost).

Operating profit is the return to all the owners of the capital used in the business. Interest is paid out of operating profit to creditors. Operating profit minus interest is net profit. Net profit minus interest is the reward to the farmer's own capital:

Operating profit – Interest paid to creditors (this is return to the lender's capital)
= Net profit (this is return to the owner's capital)

Lease payments are also a financing expense and a reward to the owners of the assets that are being leased. In such cases,

Operating profit – Interest – Lease costs = Net profit

Net profit is available to the owner of the business to pay taxes, or for consumption expenditures, or for new capital investment, or for some other use such as repaying debt. Often, an owner-operator's wage is not paid as such, and simply drawings of cash from the business are made as needed. If the owner does not actually draw out that amount and consume it, then he/she is 're-investing' back into the business some of their reward for working in and running the business, in the form of providing resources for repaying debt or new investment.

Anything left over from net profit after consumption and tax represents an increase in the owner's total wealth, called equity. The sum left over from net profit after taxes and consumption is increase in equity. It could be used to reduce a debt, or have a cash reserve in the bank, or buy an asset. For example, growth is:

Operating profit – Interest = Net profit – Income tax – Personal consumption

When calculating growth, if consumption such as drawings is less than owner-operator's salary (allowance), which has already been deducted from operating profit, then add back operator's allowance to the net profit figure, and deduct from it whatever sum is actually drawn out of the business for consumption.[2]

= Change in equity (Net worth)

Net cash flow

Net cash flow is the difference between total cash payments and cash receipts in any given period. A **cash flow budget** shows cash on hand at the start of the year, where cash came from and went to during the year, and how much was left at the end of the year. This sum becomes the cash balance for the start of the next year.

A cash flow budget can be about the past or the future—where money came from and went, or where it might come from and go to in the coming year. Cash flow budgets are used for two reasons: for planning and to check or control (**budget control**) the current year's finances through the year.

Gearing

The ratio of debt to equity is called **gearing** or leverage.

Debt ÷ Equity = Gearing

Equity

With a high debt-to-equity ratio the risk of going bankrupt is also high.

The proportion of assets (A) to debts (D) is commonly expressed as **equity percentage** (E), that is:

$$E\% = [(A - D) \div A] \times 100$$

A high equity percentage means a relatively low proportion of debt (low gearing). This means less interest and principal to repay and less vulnerability to insolvency when asset values decline. In Australian farming, equity percentage for the majority of farms is over 80 per cent.

2. Operator's allowance in the profit budget and cash drawings in the cash budget is always a point of confusion. They are not the same thing. Most owner-operators draw cash from the business as they need to, which is not the same as paying a market-equivalent remuneration to the owner-operator for the management and labour they provide to the business. Estimates of true profit have to cost all inputs at their market values, regardless of whether or not the business actually pays an equivalent cash sum for these services. So, in profit we cost owner-operator management and labour at true market value. In the net cash flow budget and calculations of change in equity, we deduct whatever cash the owner-operator draws from the business, remembering that this sum is not the same as the charge for labour and management used in the profit calculation. If the owner-operator draws out less cash than their true management and labour cost, it means the owner-operator is 'reinvesting' some of their true reward for management and labour in the business or using some of this cost as cash to use for other purposes such as to pay interest or repay principal.

Loan repayments

Loans require repayments of **principal** (the sum borrowed) as well as interest payments. The two main types of loan repayments apply to **term loans** and **amortised loans**. With a term loan, principal is repaid in equal instalments (annual, half-yearly, quarterly or monthly). Interest is paid separately on the outstanding principal and thus reduces as the principal is repaid.

Amortised loans are loans where the repayments are made at regular intervals in equal instalments consisting of both principal and interest. As the principal is repaid, the interest content of each payment is reduced, allowing more principal to be repaid. To calculate an equal annual payment, called an **annuity**, that repays an amortisation loan, use the formula for an annuity whose present value is 1, thus:

$$\text{Annuity} = \frac{P[i\,(1+i)^n]}{(1+i)^n - 1}$$

where

P = the principal sum borrowed

i = interest rate

n = number of years

For a $100,000 loan over ten years at an interest rate of 10 per cent an annuity whose present value is 1 has the **discounting factor** of 0.1627. This discount factor comes from:

$$\frac{0.1(1+0.1)^{15}}{(1+0.1)^{15} - 1} = 0.1627$$

Thus the annuity that will repay the principal and interest on a $100,000 loan is $100,000 × 0.1627 = $16,270. This annual repayment is made up of both interest and principal. In year 1 of the loan the interest component is $100,000 × 10 per cent = $10,000 and the rest, $6270, is principal. In year 2 the principal owing is $100,000 – $6270 = $93,730. So interest in year 2 is $93,730 × 10 per cent = $9373, and principal is $16,270 – $9373 = $6897, and so on.

Loan repayments should be matched to the expected repayment capacity. When a business is in financial difficulty, first look at the possibility of restructuring existing debt commitments to better fit expected annual cash surplus.

Interest-only loans are very common nowadays. An interest-only loan is a loan where the borrowed capital is not intended to be repaid on a regular and gradual basis over the life of the loan and instead annual interest is paid on the full amount of the borrowed capital for the life of the loan.

Links between profit, cash, and growth

Changes in operating profit and in annual cash surplus result in changes in the assets, liabilities and equity components of the balance sheet. A simple example of the links between a balance sheet, profit budget and cash budget is given in tables 9.4–9.7 A farm business is in the following position:

TABLE 9.4 Balance sheet at start

Assets	Liabilities
\$5 million	\$2 million 10-year term loan just started at 10% p.a. interest \$5 million assets – \$2 million debt = \$3 million equity (Equity % = \$3 million / \$5 million = 60%)
\$5 million	\$5 million

TABLE 9.5 Profit budget

Gross income	\$1.5 million
– Variable costs	\$500,000
= Total gross margin	\$1.0 million
– Overheads (including depreciation \$100,000)	\$0.5 million
= Operating profit	\$500,000

TABLE 9.6 Cash budget

Cash in	\$1.5 million
– Variable costs.	\$500,000
– Cash overheads (not depreciation)	\$400,000
– Principal repayments	\$200,000 (\$2 million × 1/10)
– Interest	\$200,000 (\$2 million × 10%)
New capital investment	\$100,000
Net cash flow	\$100,000

Growth:

Operating profit \$500,000
– Interest \$200,000
= Net profit \$300,000
– Tax 0
– Consumption 0
= Growth \$300,000

TABLE 9.7 Balance sheet at end

Assets	Liabilities
$5 million – $100,000 depreciation + 100,000 cash from NCF surplus + 100,000 new capital investment = $5.1 million	$2 million – $200,000 principal repaid = $1.8 million $5.1 million – $1.8 million = $3.3 million equity Equity % = $3.3 million /$5.2 million = 65%

What has happened?

- Assets have increased by $100,000 ($100,000 new investment out of NCF, plus $100,000 surplus NCF minus $100,000 depreciation)
- Debt has reduced by $200,000 (principal repaid on loan)
- Equity (A – L) has increased by $300,000, from $3 million to $3.3 million, growth of 10 per cent
- Equity % has increased from 60 per cent to 65 per cent (A – L)/A.

Comparing businesses

Advisers to businesses who do not have economics as the **core discipline** in their analyses often use an approach to analysing businesses known as **comparative analysis** or benchmarking. The essence of this approach is the belief that it is possible to learn much about what to do in one business by having a close look at the average performance of other businesses engaged in similar activities.

Comparing and contrasting one business with another can provide useful insights about different technical processes to achieve particular ends. However, comparative analysis and benchmarking has many limitations in identifying problems and the true cause-and-effect relationships within a business.

All businesses comprise a unique mix of resources. First, the people operating the businesses have different goals, skills, attitudes, experiences, family situation and stage of life. Second, the quantity and quality of resources available to different farm businesses are different. Third, farming is different to most other businesses in the exposure to volatility and lack of control over key determinants of outcomes. This means measures can vary greatly from year to year regardless of how the business is managed.

Superficially, it seems useful to look at one farm that produces say an average of 100 kg of wool per hectare, or an average of 5000 litres of milk per dairy cow, or carries on average 2 cows to the hectare, and then compare these figures with what happens on another farm. However, this approach is theoretically flawed. Maximising average technical ratios of output to input is not the means to maximise profit. Indeed, relying on this sort of information can lead to drawing logically opposite conclusions about what to do. For example, maximising the ratio of milk per cow would involve reducing the number of cows per hectare and increasing the milk per head; maximising the ratio of milk per hectare would involve increasing the number of cows per hectare and reducing the amount of milk produced per head. These are logically opposite conclusions.

The correct analysis of how to increase or maximise profit is to have a stocking rate of cows per hectare and a level of production per cow at which the **marginal cost** of producing an extra unit of output just equals the marginal return from an extra unit of output. The focus of the analysis and the decision is on expected extra return and extra cost from a change to the whole input: output make-up of the system, not an average of one technical component.

If benchmarking of technical productivity ratios between different systems involving different people with different resources, goals, attitudes to risk, stage of life and so on, is not a meaningful exercise, how useful is comparing measures of economic and financial performance between different businesses?

Production, costs, income, capital and debt of a business lend themselves to the formation of ratios that summarise aspects of the business. These 'summary ratios' are used to compare with similar other 'summaries' about businesses. For ratio analysis to be meaningful, the measures need to be meaningful. Crude aggregate measures, such as using gross sales or output as some proxy indicator for profit or net cash flow or growth, are not as informative as summary ratios that account for both the income and the costs. Agriculture is so volatile that gross measures can be associated with good or bad net results!

The gearing (or leverage) ratio measures the relationship between debt and equity of the farm business. The higher this ratio the larger are the outside claims on the business relative to the equity of the business. Lenders have a particular interest in this gearing ratio as it is one measure of the financial risk involved. Another ratio of interest is the ratio of fixed cash cost to total cash costs. The higher the value of the fixed expense to total expense ratio the less flexibility the farm operator has to adjust quickly and efficiently to changing market conditions.

The major limitation of financial ratios is that they are historical averages, whereas management, and economic theory, is about marginal concepts and the future. Economic theory tells that if firms wish to maximise profits they should operate at the point where marginal cost is expected to equal **marginal revenue**. As ratios are based on averages they can easily be misinterpreted. Businesses could have the same rate of return on capital because they have the same average costs in producing output and they sell at the same price. However, one operator could have a low rate of return compared to the potential return because the size of the business is too small and not producing enough, and another operator could have a low rate of return because the business is too large and is operating inefficiently.

While there is no limit to the summary ratio measures that can be, and usually are, produced about a business, only a few are really useful in telling the owner much of value about the business. The main ones are measures of efficiency (for example operating profit as a percentage of total capital); liquidity (for example net cash flow before and after debt servicing, interest rate coverage ratio); growth (for example change in equity, gearing ratio) and financial risk (gearing ratio). Trends in such measures over the recent past for a particular business are informative. More useful are the expected movements in the important measures over future planning periods. Comparison of actual and expected business performance, especially cash flows, is also very useful information.

To sum up: the focus on historical average performance is not useful to management that is concerned with future changes and the extra benefits and extra costs of the change. It is not possible to deduce the marginal consequences of a change to a farm system from historical average measures of performance, let alone deduce these for one business by having a good hard look at another business. Further, a manager achieving what looks like poor results, could be doing a terrific job with terrible resources; another manager in another situation might look to be achieving good results and might be actually doing a terrible job with terrific resources.

Comparisons of average benchmark measures of performance between different farm systems—league tables of top, middle, bottom, best, worst and so on—tell little about what is actually being achieved and nothing about why it is so. There is an implied cause and effect between a measured average phenomenon and an observed set of actions that may or may not be the true explanation of cause and effect. The further implication is that if some other business does similar things to the benchmark system, similar results too will be achieved. This is not usually so.

The perennial farmer response to benchmarking data is a querulous: 'now what do we do?' The only way to determine true cause-and-effect relationships and solve business problems is to know well the business and the people in it, walk the farm, understand the technology and the technical possibilities, and have mastery of the economics, finance and risk analyses. The only way to provide sensible answers to the question 'and now what do we do?' is to properly analyse the expected benefits and costs of feasible potential changes to the system. This is a process that is way beyond benchmarking. Above all, the essence of successful farming is implementing new technology in changed systems. There are no benchmarks for innovations.

Analysing change on farms

Adapting the business to changes is the core of the management task. Changes to farm plans can be a quick change in an aspect of an activity, or can involve significant capital investment that takes time to reach full fruition.

Market considerations

The first question to ask when considering any changes to the farm operation is 'what are the market prospects for the output from the change?' Farm product has to be sold at a price that will cover costs and leave some profit. The first step in the process of changing resource use on a farm is to weigh up the market prospects for the particular output involved, in terms of quantity and quality required, and prices that might be received.

Human aspects

The farmer and the employees need to have the interests, skills and knowledge, and the right incentives to make a change succeed. The appropriate supply of labour must be available when needed. The stages of life of the members of the farm family—starting out, children being educated, expanding, winding down and so on—are important in determining what changes are appropriate for a business at any particular time.

Technical knowledge

A sound knowledge of the technical aspects of a proposed change is necessary. The decision maker needs to investigate when and how to sell the product(s) resulting from the change and the likely supply of and demand for the product. The physical aspects to consider are whether the resources are available (labour, soil type, water, animals, pasture, machinery) to carry out the change, or whether extra resources will have to be obtained.

Technical needs may include specific types of fertiliser, specific animal husbandry techniques and the methods of growing, harvesting and marketing. A detailed physical plan of the land, machinery, crops and types of animals is required. The technical basis of farm budgets has to be correct.

The farmer's detailed knowledge and experience of production on the farm is the main source of the detail that is needed. Experimental results from research plots have to be adjusted to the specific situation of each farm. Yields used in budget calculations have to be those that are possible using economic levels of inputs on the particular farm.

Comparing profit from changed farm plans

One method of evaluating whether it is worth making a change is to compare farm profit for a year of operating with the change and farm profit for a year without the change. The technique of comparing with-change and without-change situations can involve comparing two whole-farm budgets, one with and one without the change, or a doing a partial budget, which looks at the effects of the change only.

Regardless of the nature of the change, impacts have to be assessed in terms of the effects on the whole farm. The economic way of thinking dictates that the perspective to use when considering a change is 'what might be the situation if the change is not made?' compared to 'what might be the situation if the change is made?', and then 'which situation is the one I would prefer to face?'

An approach to evaluating the merit of a change is shown below, using a series of questions that need to be answered.

- What is expected annual operating profit of the farm as it currently operates—without change?
- What is expected annual operating profit as a percentage of total capital (return on total capital) of the farm as it currently operates?
- What is expected annual net cash flow of the farm before and after debt servicing as it currently operates?
- What is expected annual operating profit of the farm with change?
- What is expected annual operating profit as a percentage of total capital (return on total capital) of the farm with change?
- What is expected annual net cash flow of the farm before and after debt servicing with change?
- Is the (1) extra profit, (2) extra return on extra capital and (3) extra net cash flow compared to extra debt servicing requirement, satisfactory?
- What other things will change but are not accounted for well in the analysis so far? Do these factors change the conclusion about whether or not to implement the change?

Example:

A farm currently has $4 million in total capital made up of land, livestock and machinery. It is expected to earn $200,000 per year: a return on total capital of 5 per cent p.a. before tax. It has $500,000 debt as a ten-year term loan, just started and so in year 1 the interest charge is 10 per cent p.a.

Annual debt servicing requirements are $50,000 for principal repayment and $50,000 for interest payment in year 1. The farm has an expected annual cash surplus of $150,000 before debt servicing. This more than covers the debt servicing needs.

A potential change is identified. This will involve replacing an animal activity with a cropping activity. The change will be in the activity gross margins that make up whole farm total gross margin, and there will be some extra cost from depreciation on the extra cropping equipment that has to be purchased.

The analysis for a one-year budget to evaluate a change can be framed as shown in table 9.8.

TABLE 9.8 Analysis for a one-year budget to evaluate a change

Current activity to be given up	New activity to be implemented
Annual gross margin	Expected annual gross margin
	Minus extra annual depreciation on extra capital and other extra annual fixed costs
(A) Current annual benefit forgone if change made	(B) Annual net benefit from new activity if change made

(A – B) = Annual net benefit from making the change, before extra tax, in dollars

This sum 'Annual net benefit from making the change' can be expressed as a percentage return on the extra capital invested in changing to the new activity. That is (Net benefit/Extra capital)*100/1. In addition, other non-financial benefits and costs have to be identified and considered.

In this example, the livestock activity that is being replaced currently produces an annual gross margin of $5000 p.a. This will be given up if the change is made. The cropping activity is expected to produce an annual gross margin of $40,000 p.a. Cropping activity requires $100,000 in capital for equipment. The extra annual depreciation of cropping machinery is $10,000.

The annual net benefit from making the change is expected to be: the extra gross margin minus extra depreciation from the cropping activity, minus the gross margin forgone from the livestock activity: $40,000 – $10,000 – $5000 = $25,000 per year.

The gain of $25,000 per year represents a return on the extra capital invested of $25,000/$100,000 = 25 per cent before extra tax. This annual return on extra capital is attractive as it would increase total farm profit by $40,000 – $10,000 – $5000 = $25,000. Annual operating profit of whole farm total capital becomes $200,000 + $25,000 = $225,000.

The total capital of the farm after the change would be $4 million originally invested minus $20,000 capital no longer invested in livestock plus $100,000 new machinery capital = $4,080,000. The percentage return on total capital after the change would be $225,000/$4,080,000 = 5.5 per cent p.a.

So, what has happened? Is the change worth doing? The answer is yes, for two reasons. First, the return on the extra capital is expected to be 25 per cent, which is more than a similar amount of capital could earn if invested elsewhere for a year in similarly risky ways. The farm business is making more profit and better returns to capital with the change than was the case without the change.

On the cash side, the livestock activity contributed the gross margin of $5000 to whole farm cash surplus and the cropping activity is expected to have a cash surplus of $40,000 (before non-cash cost of depreciation). Even if a cash sum equivalent to annual depreciation of $10,000 is invested to replace in the future the extra machinery that was purchased for cropping, there is $30,000 cash available after the change to service the extra borrowings associated with the change.

If the extra $100,000 required for machinery was made up of $20,000 from the livestock capital freed up and a ten-year term loan for $80,000 borrowed at 10 per cent p.a., the $80,000 new debt would require $8000 principal repayments in year 1 and $8000 interest repayments. The extra $30,000 cash from the change will cover the extra debt servicing. The previous farm plan also had some cash surplus, which means extra **debt-servicing capacity** too if needed.

The risk associated with the proposed change has to be included in the evaluation of extra profits. Many apparently good ideas are not adopted because once risk is taken into account the expected returns from the project are not high enough to cover the risk that it might fail. In the example given above, the extra return on capital of 25 per cent, and the surplus cash after debt servicing, combine to make the investment attractive compared to alternative investments of the $100,000 capital involved.

The one-year budget approach is one method used to evaluate change to a farm plan, even though the change will go on for a number of years. If the change takes several years to reach full operating capacity—called steady state—then the one year looked at should be when the change is fully operational, in the steady state.

The most accurate method to evaluate the return on capital of a change that spans a number of years is the discounted cash flow method. This method is more complicated than the single-year method, and is not dealt with in this introductory text (for more on this approach, see Malcolm et al. 2005).

Risk and agricultural production

Business and financial risk

Risk is a term given various meanings, but it always relates to the volatility of potential outcomes. Risk is conventionally classified into two types: *business risk* and *financial risk.*

Business risk is the risk any business faces regardless of the amount of debt and equity and level of interest rates. It comes from production and price risk. Business risk refers to variable yields of crops, reproduction rates, disease outbreaks, climatic variability, unexpected changes in markets and prices, changes in government policies and laws, fluctuations in inflation and interest rates, and personal mishaps. When yields, prices and costs are expected and do not happen, it reduces the ability of the business to pay for the inputs used, to service debt and to appropriately reward labour, management and capital.

Financial risk derives from the proportion of other people's money used in the business, relative to the proportion of owner-operator's capital, and the level of interest rates. The higher the ratio of debt to equity, the higher the gearing ratio, the higher is the financial risk faced by the business because of the operation of the principle of increasing risk. Financial risk exacerbates business risk.

Good decisions

The difference between a good decision and a right decision is fundamental to understanding the role of risk in farm management decision-making. A good decision is based on the best information and judgment available at the time the decision is made.

A useful practical approach in assessing risk is to test the sensitivity of results to changes in critical variables, using the 'what if' approach. For example, if I do 'this', and 'this' happens, then the ultimate outcome(s) will be like 'this'. The decision maker weighs up all the information in the light of their own judgment about how likely it is that the various important events and outcomes could happen.

Estimating **break-even** levels of key **parameters** is useful in **decision analysis**. 'Break-even' is the level of a key parameter(s) that needs to prevail for the decision maker to be as well off after the change under investigation as before the change.

In practice, farmers take many different steps to place their business in a 'risk-situation' that gives a good chance of long-term survival. These include (among other things):

- being good at the technology (this is one of the most important steps)
- ensuring that the business is not highly geared
- keeping overheads low relative to output, gross income and TGM. This is achieved by keeping production up, judicious expansion, prudent investment in machinery and stock, and not having too many family members trying to gain a living from the farm
- specialising or diversifying (either can be the best risk strategy, depending on the case at hand)
- having flexibility in the business structure and the farming system
- building up non-farm assets when it is possible; and earning a reasonable and probably increasing amount of net income and net cash flow from sources off the farm. Depending on the state of development of the farm, the investment with the highest return after tax on marginal capital could be on the farm or off the farm. Beyond some reasonable level of farm

productivity, sound investment off the farm is prudent. However, it has to be in something the investor understands.

REVIEW QUESTIONS

1. Health of a farm business is assessed in terms of efficiency (profit), growth (wealth) and liquidity (cash). Explain how profit, growth and cash are measured in a farm business.
2. The outstanding characteristic of a farm business is the need to manage risk. What are the main risks farmers deal with and how do they manage them?
3. You can only validly compare yourself with yourself. Explain why it is more useful to compare the performance of a farmer's business over time than comparing the performance of a farmer's business with other farmers' businesses at any one time.
4. Drought is a normal part of farming in Australia. What can farmers do to manage drought?

FURTHER READING

Blackburn, A. and Ashby, R., *Financing your Farm: A Practical Guide to Financial Growth* (4th edn), Australian Bankers Association, 2006.

Boehlje, M. and Eidman, V., *Farm Management*, John Wiley and Sons, New York, 1984.

Malcolm, B., Makeham, J. and Wright, V., *The Farming Game: Agricultural Management and Marketing*, Cambridge University Press, Cambridge, 2005.

WEBSITES

Australasian Agribusiness Review and Australasian Agribusiness Perspectives: www.agrifood.info/

Australasian Farm Business Management Network: www.csu.edu.au/faculty/science/saws/afbmnetwork/

Australian Agricultural and Resource Economics Society: www.aares.info/

Australian Bureau of Agricultural and Resource Economics: www.abareconomics.com/

Australian Farm Institute: www.farminstitute.org.au/

CSIRO: www.csiro.au/

National Farmers Federation: www.nff.org.au/

NSW Department of Primary Industry: www.dpi.nsw.gov.au/

Vic Department of Primary Industry: www.dpi.vic.gov.au/dpi/index.htm

PART 6

Changing Agriculture

Sustaining Agriculture

10

CONTENTS

Introduction

Farming is mostly about managing resources in the face of great uncertainty about the situation being managed. Farmers have to deal with short-term fluctuations in weather and markets as well as prolonged changes in the natural, economic and social conditions in which they work. Since the mid 1990s most farmers across Australia have experienced drier seasons than they had experienced, though not any drier than had happened at various times since European settlement.

Agriculture is constantly changing. A significant recent change is the demand by society for people to reduce carbon pollution and for farmers to supply environmental services. The extent of demand by society for people to reduce carbon pollution and for environmental services such as biodiversity and landscape from landowners (predominantly agricultural enterprises) is without precedent. A corollary of this change is the increase in demand by society to minimise or eliminate external costs arising from agricultural enterprises (e.g. carbon additions to the atmosphere, and nutrient loss from farms to the environment).

By most standards of international comparisons, Australia is a dry country in terms of total annual rainfall recorded in the populated areas. In all likelihood

Australia's farmers for the next several generations will have to become better at dryland farming than they have been in the past, as temperatures rise and rainfall declines and becomes more volatile. Drier and more volatile natural conditions look like being the lot of farmers in south-eastern Australia and south-western Australia in the coming decades.

As always, farmers will adapt and continue to farm profitably in the changed conditions, because change creates challenges and challenges create opportunities, albeit only for some but not for all. As always the race will be won by the most fit. The most fit farm businesses will be those that have the best opportunity to adapt and exploit the drier and hotter conditions of production and the market opportunities that will arise as farmers around the whole world encounter changes in climate.

The keys to prospering in the changed climatic and market conditions of the future will be:

- being technically good at farming, as has always been the case
- growing the size of the business over time to continue to reap the benefits of economies of size as technical and economic conditions change
- having flexibility to capitalise on opportunities when they arise, which stems from having a mix of overhead and variable costs that enables strategic intensification and extensification at different times in different circumstances
- diversifying spatially, by activity on and off farm to harness the benefits of the portfolio effect in managing risk. For farming, the portfolio effect in managing risk will involve reducing the exposure of the farm business to any particular risk by investing in activities that are not affected by the same source of risk at the same time, in different places and on and off farm.

Sustainability

Sound farm management aims to ensure that resources can continue to be productive in the future: that is, resource uses in farming are sustainable. There are many definitions of the term *sustainability*. Meaningful definitions of sustainable agriculture have both technical and economic dimensions. In focusing on the use of resources it is worth keeping in mind that (1) resources can be substituted in the process of production, and (2) over time, new technology alters the resources, and the mix of resources is needed to produce the goods and services people want.

Since European occupation the productive capacity of Australian farmland has changed markedly: mostly for the better through pasture improvement and more productive cropping practices. The extent of the problem of the quality of farmland and water supplies deteriorating has to be judged according to measured criteria. Assessing degradation of resources requires accurate measures of the current conditions of resources throughout Australia. It is also helpful to have accurate indications of the previous situation, and to ponder what would have been the current situation if agricultural resources had been used differently, known as the 'compared with what?' question.

Nevertheless, the fact remains: the quality of water supplies and of some farmland has deteriorated. Why has this degradation occurred? Either the land is unsuitable for the production activities that have been undertaken or the farming practices themselves have been the problem. Either way, when degradation of resources occurs, it is in the interests of farmers and the wider community to develop and adopt less degrading practices.

An imperative of Australia's farm sector remaining profitable is increasing productivity. This needs to be achieved in ways that maintain the productive capacity of the farm resource without causing irreparable damage to resources either on or off the farm. Always there is the need to understand the impact of farming practices on the overall quality of resources that will be available for use in the future.

There are many examples of practices that were 'thought to be sustainable' but have proved to be inherently damaging: for example the soil-acidifying effects of annual legume pastures; the onset of herbicide resistance in continuous grain legume / cereal rotations in WA; and dryland salinity resulting from too much clearing of land off deep-rooted plants and trees. Close monitoring of emerging production systems has to be maintained to identify 'any emerging biological defects' in the farm practices.

What does sustainability mean?

The United Nations World Commission on Environment and Development (WCED, the Brundtland report, 1987) defined sustainable development as 'development that meets the needs of the present without compromising the ability of future generations to meet their own needs'.

In Australia the ideal of sustainability also includes the well-being of the environment. The federal government Working Group on Sustainable Agriculture in 1991 found that the essential feature of sustainable development (including agriculture) is recognition that (while in some cases there will be a trade-off between economic growth and the environment) economic management can have a positive impact on the environment and improved environmental quality can enhance the performance of the economy. This means that if the environment is looked after then society can be better off in both economic and environmental terms.

It is not simplistically the case that productive activities make good economic sense and bad environmental sense: meeting economic and environmental imperatives can be complementary aims. The relationship between economic growth and development and the state of the natural environment is positive: the wealthier the country, the better the state of the environment.

With these views in mind the Working Group on Sustainable Development endorsed the following definition of sustainable agriculture: 'the use of farming practices and systems which maintain or enhance the economic viability of agricultural production, the natural resource base, and other ecosystems which are influenced by agricultural activities'.

The Working Group proposed a set of principles for sustainable agriculture, described as 'broad goals against which all kinds of policies at all levels of government could be judged'. The principles listed below deal with issues of farm viability (principles 1 and 5), environmental damage (principles 2 and 3) and the benefits to society at large (principle 4). These principles are useful for assessing whether or not a particular farming system is considered sustainable farming.

Principle 1 Farm productivity is sustained or enhanced over the long term.

Principle 2 Adverse impacts on the natural resource base of agriculture and associated ecosystems are ameliorated, minimised or avoided.

Principle 3 Residues resulting from the use of chemicals in agriculture are minimised.

Principle 4 The net social benefit from agriculture is maximised.

Principle 5 Farming systems are sufficiently flexible to manage risks associated with the vagaries of climate and markets.

Sustaining the state of natural resources

The scientific view of the state of natural resources favours the idea that changes in the condition of natural resources from one state to another are something that has to be fixed or restored. In contrast the economic view is that if the change has happened, then it is a cost that has been incurred—regrettable but a sunk cost. The question becomes: 'what are the benefits of fixing the situation and what are the costs?' If the benefits exceed the costs, then fixing things will make people better off and it should be done. If costs of repair exceed the benefits, then the fixing should not be done and the resources that would have been used to fix the situation would add more to the welfare of people if they were used in some other way.

It follows from the economic view of natural resource use and conservation that there is an optimal amount of resource degradation. This is the amount at which the cost of reducing degradation by one more unit will add less benefit than the cost of doing it. Up to this level of resource degradation, the benefits of reducing degradation have exceeded the cost of doing it and so should be done. It is rarely economical to remove the last weed from a crop of wheat!

Farmers are at times depicted as being users of natural resources such as land who inevitably reduce the natural qualities of that land. In agriculture, land represents the major part of the farmer's wealth, and farmers have strong incentives to protect the value of their wealth. Their profit each year depends on how productive their land is. Farmers have strong motivations to protect and improve the productive quality of their land.

Soil degradation

Erosion

Soil erosion is the loss of surface soil from a paddock by water run-off or by strong winds; both wind and water dislodge small soil particles and carry them away. Erosion causes topsoil to move from one farm to another. Topsoil is the most fertile soil layer. It stores much of the water and nutrients used by crops and pastures. Loss of topsoil by erosion will reduce short- and long-term production. For example, the removal of 1 cm of topsoil (the equivalent of 120 tonnes of topsoil per hectare) from good wheat-growing paddocks to somewhere else has been estimated to reduce wheat yields by an average of 150 kg per hectare (which is around 10 per cent of the national average wheat yield), by removing essential plant nutrients that are concentrated in the topsoil. The alternative way to see this is that erosion imposes costs on farm businesses. Erosion that removes soil and nutrients means that farmers have to spend more on fertiliser to replace soil nutrients that have been blown away or washed away, and there is also a degradation of soil structure. The problem of soil erosion can be widespread in the cultivated areas in dry years, and in all areas in very wet years. A key economic aspect of soil erosion is that the effects are mostly on the farm, with relatively few off-site effects. This means that the costs of erosion are borne mostly by the farmer.

Water erosion increases as the rate of rainfall run-off (the rain that cannot infiltrate into the soil) increases, and is determined by the extent to which the surface soil is able to be removed by the run-off water. High intensity rainfall, steep slopes, and finely cultivated surface soil with no plant cover are a recipe for massive soil losses by water erosion. Such losses are often associated with a few catastrophic events, where soil losses would be around 50 tonnes per hectare for that event. Water erosion is preventable, or at least can be reduced to very low levels. Steep land should not be cultivated. Sloping land that is cultivated needs to be protected by installing contour banks to reduce water run-off across the slope.

Another approach is to undertake strip cropping. This involves growing crops in strips across a gently sloping paddock as is done in areas of Queensland and northern NSW where heavy summer rains fall on summer fallows of fine self-mulching clay soils. Strip-cropping means that there will be strips of fallow, crop or stubble at any time across the slope and these will retard the rate of movement of water over the fallow strips, thereby reducing run-off and increasing infiltration into the soil. Plant cover has to be maintained on the soil surface by retaining crop residues on cropped land or by avoiding overgrazing, which will expose the soil surface on sloping pasture land, particularly if rainfall intensity is likely to be high.

Wind erosion is the major source of erosion in the drier parts of southern Australia. It results when strong winds blow across exposed sandy soils that have minimal vegetation cover. Sandy soils have wind erosion rates perhaps a hundred times that of clay soils because the proportion of soil aggregates below a critical size (around 0.85 mm diameter) is much higher in a sandy soil. Cultivating sandy soils greatly increases the risk of wind erosion.

The key to minimising wind erosion on vulnerable soils is to protect the surface with retained stubble or pasture, or even a carpet of weeds. It is necessary to retain around 1 to 2 tonnes of crop stubble per hectare for effective protection. Ideally, to reduce the risk of serious soil erosion, cultivation should be carried out as late as possible prior to sowing the next crop to minimise the period of exposure. Thus, the retention of ground cover is the key to preventing soil erosion. This has implications for tillage practices, stocking rates and for whole-farming systems.

The main message for farming systems is that there has to be flexibility in the system to enable the farmer to respond to the weather conditions that predispose the soil to erosion. When the expected rains do not arrive, there is a danger of the ground remaining bare during the period when erosion risk is highest. This could happen because it is too dry to sow the crop, or the land might become bare through overgrazing.

Acidification

There are over 25 million hectares of agricultural land in Australia where the soils are strongly acidic, with a soil pH_{water} of less than 5.5. These soils are not particularly fertile in their native state. As the soils become strongly acidic the supply of phosphorus, calcium, magnesium and molybdenum become less available and toxicities of aluminium and manganese can develop, particularly in the more acid soils. The combined effect of deficiencies and toxicities limits the production of a wide range of agriculturally important plants.

Soil acidification is a natural process that occurs as soils undergo further weathering and lose nutrients through leaching. A number of farming practices, such as the use of productive legume-based pastures or the use of nitrogen fertilisers, accelerate the soil acidification process. They cause an increase in the oxidation of nitrogen (as ammonium) to form nitrate, a process that produces acid (H^+ ions). If this nitrate leaches from the soil, as happens in a number of annual pasture systems, macronutrient cations (Ca^{2+}, Mg^{2+} or K^+) will accompany the nitrate ions. The increase in plant production in the short term, associated with the higher nitrogen status, increases soil organic matter and the cation exchange capacity of the soil. The net effect is that there is a shortage of these cations to balance additional organic anions and the extra negatively charged sites on the organic matter. The hydrogen ion concentration increases to correct this imbalance.

Other farm practices such as haymaking also accelerate loss of these cations from the soil and speed up the rate of soil acidification. However, soils differ in the rate they can acidify, so the urgency to address acidity problems varies. Soils with a light texture (sands to sandy loams with low clay content) and low organic matter content have a low pH buffer capacity. This means the soil does not have the capacity to resist or prevent a fall in soil pH as acid is produced in the soil.

Applying finely ground limestone ($CaCO_3$) is the most practical way to neutralise acidity in a soil. The amount required to regain yield potential in an acid soil varies from 1 to 6 tonnes per hectare, depending on the desired increase in soil pH and on the pH buffer capacity of the soil. Liming will not prevent acidification, and the pH of limed soil can still fall in time. Applying lime is costly. For many producers the profitable choice is to grow acid-tolerant species (e.g. cocksfoot, triticale, lupins) on acid land. This can maintain profitability in the short and medium term, but acidification will continue in the long term. Eventually a neutralising product such as lime will be needed to maintain production.

Salinisation

Soils become saline when salts (mainly NaCl) accumulate in the root zone. The build-up of salts kills the existing vegetation because they are toxic to the plants and the plants are unable to use the soil water due to the osmotic effect of the high salt concentrations.

Another effect of salt accumulation is on the soil itself. This causes the soil aggregates to disperse when wet, which reduces aeration and water infiltration into the soil and leads to the formation of a hard surface crust. The saline areas tend to be lower than the surrounding land, ending up as bare patches of erosion-prone, unproductive and frequently water-logged soil. The only plants that will grow on them are unproductive salt-tolerant species.

Salts in salinised soil come from rising underground watertables. As the watertable rises it brings with it dissolved salts from deeper subsurface layers. These salts have accumulated from the weathering of rocks, or from rainfall, or from the salts that have remained in the soil after sea waters have receded in past geological times. As the rising watertable containing dissolved salt approaches the soil surface, moisture moves to the soil surface by capillary action. The water evaporates at the surface. This leads to the build-up of dissolved salts in the surface layers. Further rises in the watertable cause saline water to seep onto the soil surface. The damage is done: the productive capacity of that soil is lost, unless the watertable can be lowered and the salt leached out of the soil.

There are two types of soil salinity. They differ according to the causes of the rising watertables. The first is *dryland salinity*. This has resulted from the clearing of native perennial vegetation, particularly woodlands and forests, to grow annual crops and pastures. The annual plants use less water so more rainwater drains through the soil profile below the root zone and enters the groundwater, causing the watertable to rise.

The rising watertable discharges occur many decades after the land has been cleared. The worst areas for dryland salinity are in Western Australia and in Victoria but the problem is increasing in South Australia, NSW and in Queensland. Increasing the proportion of deep-rooted perennial plants in the landscape is one solution to the problem of dryland salinity.

The second type of soil salinity is *irrigation salinity*. In Australia this occurs on parts of the riverine plain of the Murray and Murrumbidgee Rivers in Victoria and NSW. In addition to increases in the groundwater levels that result from clearing native vegetation, groundwater increases from the drainage of excess irrigation water below the root zone.

It is difficult to irrigate and not have some of the water drain below the root zone. The amount of extra drainage might be small, but with frequent irrigations during the summer, the cumulative amount will be significant. Solutions to this problem are: more precise applications of irrigation water according to soil moisture and plant measurement; running water on and off land at the optimum speed to achieve the right penetration and lateral movement; sub-surface drainage and re-use catchments of water; and the disposal of the surplus water, which frequently becomes saline. Unfortunately it is difficult and expensive to effectively dispose of surplus saline water. Irrigation schemes around the world and throughout history have failed because of the scourge of irrigation salinity.

FIGURE 10.1 Soil salinisation processes

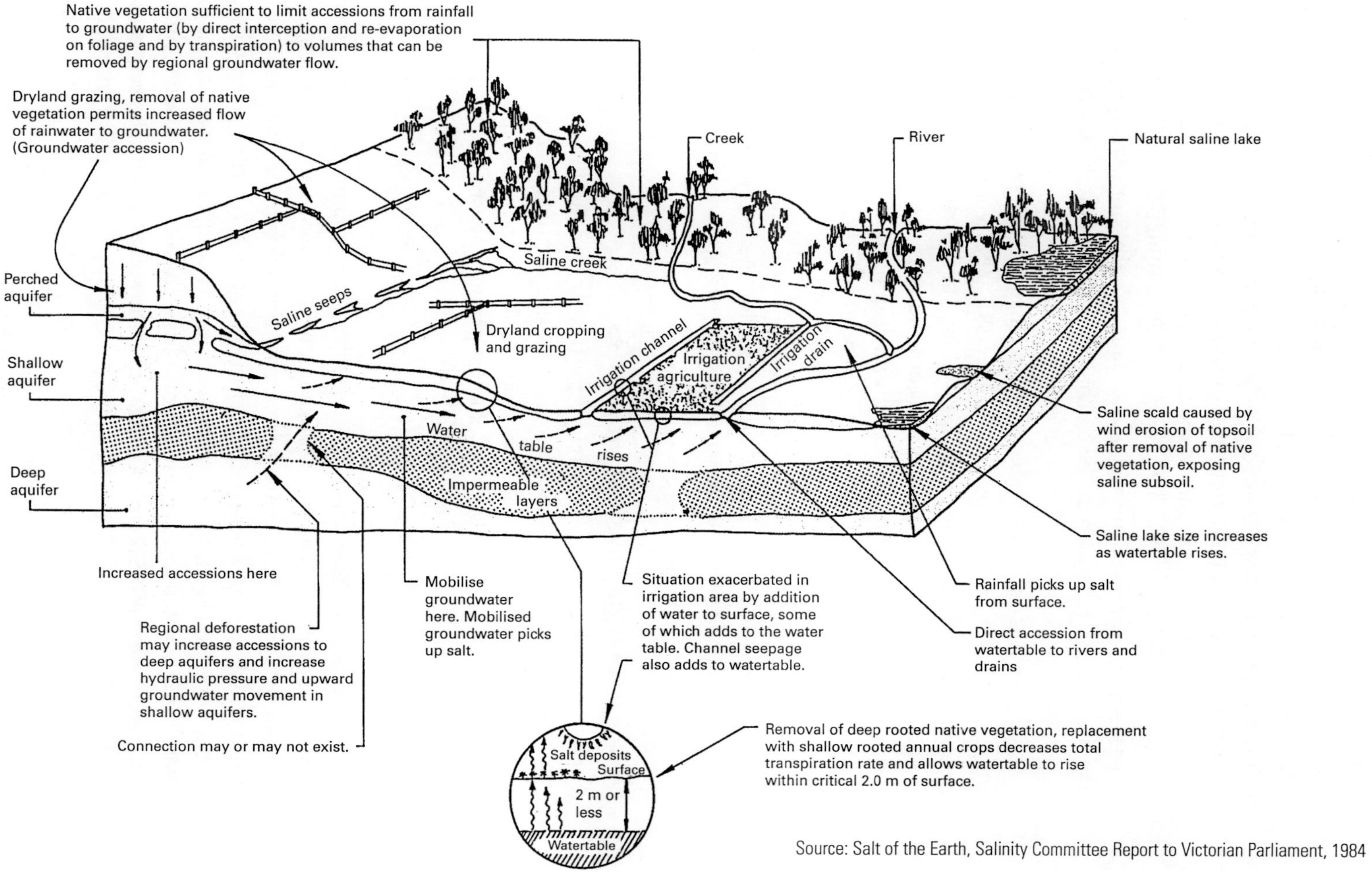

Source: Salt of the Earth, Salinity Committee Report to Victorian Parliament, 1984

The solution to irrigation salinity is to correct the hydrological imbalance that has occurred. This means preventing surplus water from entering the groundwater in order to avoid further rises in the watertable. This is difficult to implement: it involves reducing water recharge into the groundwater and increasing the removal of water from the groundwater in discharge areas.

In the case of dryland salinity, reducing salinity involves minimising water additions in recharge areas by planting deeper-rooted perennial vegetation that transpires more of the water that falls as rain. Similar high-water-use plants could also be grown at the discharge end of the landscape to lower the watertable.

Salinity in Australia is a problem that ostensibly affects a small area of agricultural land. But, unlike erosion, the effects of salinity are not borne only by the irrigator who is contributing to the problem, or the pastoralist whose grandparents cleared the land of timber and contributed to the problem, or the grain farmer whose cropping practices result in deep drainage and water recharge. The salinity problem is frequently experienced by other farmers whose farms are lower in the landscape or who are downstream from the irrigation area, and by non-farmer users of water whose water quality has deteriorated because of excessive irrigation and salinity. The corrective measures involve more than just actions by the farmer who is applying the irrigation water, but also involve the community at large.

Deterioration of physical properties

A soil that has favourable physical properties for root growth has a soil structure that consists of distinct soil aggregates that are held together by organic material. The aggregates and the pore spaces between them allow for optimal infiltration, storage and release of water for plant uptake and for the movement of air through the soil mass. A 'well-structured' soil provides a favourable medium for root growth and contributes to the vigorous growth of crop or pasture plants.

The physical condition of many Australian cropping soils has deteriorated. Soil aggregates have been broken down by excessive cultivation and the repeated passage of farm machinery across the soil surface. The outcome is a compacted dense soil: the surface layer may dry to a hard crust, or the whole of the cultivated soil layer may set hard on drying. A compacted layer might form below the cultivated layer. Compaction of the soil slows down water infiltration, reduces the movement of air through the soil and restricts root growth. Plant growth is less in such soils, so crop and pasture productivity declines.

The decline in the surface structure of cropping soils goes hand-in-hand with a decline in soil organic matter. The stability of soil aggregates depends on organic materials binding finer soil particles together; repeated tillage leads to a breakdown in these bonds and a loss of the organic material.

FIGURE 10.2 A compacted plough pan at a depth of 8 to 15 cm in a red earth soil in northern NSW

Physical damage to soils is more apparent in lighter-textured soils such as the solodics, the red and red brown earths, and the podzolics, such as those that make up extensive areas of the southern and Mediterranean cropping belts. The cracking clays that predominate in the northern cropping areas have a shorter cropping history but are now starting to show signs of physical deterioration.

Nevertheless, farm practices change. Management practices that reverse deterioration in soil structure involve improving organic matter levels in the soil. The use of legume pasture phases in cropping rotations has been effective in the past. An approach that has been widely adopted has been direct drilling and chemical fallows. These methods reduce the amount of soil cultivation. Combining this with stubble retention improves the chances that the soil physical condition will be maintained.

Depletion of nutrients

There is evidence that nutrient depletion has occurred and is continuing to occur on many cropping and pasture soils in Australia. Nutrient depletion is where nutrients in the soil decline unless increasing nutrients are added. Around 20 kg of nitrogen is removed per hectare for every tonne of wheat grain that is harvested. Marked yield responses to N fertiliser begin to occur as soil N declines. In the southern wheatbelt, soil N levels in cropping land, which previously were maintained by vigorous pasture legume phases, declined as the productivity and legume content of the pasture phase declined. The use of grain legumes in the rotation plus N fertiliser is needed to produce profitable grain yields.

Another problem of nutrient depletion is potassium (K) being removed from grazed pastures. This problem developed on light-textured soils in Western Australia where improved pastures had been used for more than fifty years. Emphasis was on supplying P and S in single superphosphate as adequate P was essential for legume growth. Initially potassium was not an issue. Yet the K status of sandy soils is low, and K was removed in animal product. Once K levels are depleted, K fertiliser is required to maintain productivity and profitability.

The lessons from these examples are clear. Cropping and grazing activities remove mineral nutrients from the soil. Nutrients are added naturally from the weathering of rocks, from the atmosphere, and in the case of N, by biological fixation. The rates at which nutrients are added by these natural processes are often too low to sustain desired production levels. Higher levels of productivity require additional nutrient inputs to the farm system over time.

Vegetation decline

Loss of productive species and weed invasion

There are numerous examples of loss of desirable pasture species from higher-rainfall pastures or from the drier rangelands. These changes have invariably resulted in the replacement of the desirable species by more 'weedy', less productive, plants. The net effect has been a decline in the productive capacity of the pasture or rangeland. The delicate ecological balances that exist in the rangeland ecosystem were not understood.

An example of degradation in pastoral areas has been described for the upper catchment of the Burdekin River in north Queensland. Considerable increases in beef cattle numbers in this country resulted from the switch to the better-adapted *Bos indicus* cross-bred cattle (that could survive the dry years), the supplementation of dry feed with non-protein N and mineral supplements, and the use of road transport to enable stock to remain longer on overgrazed pasture (in droughts) and return from agistment sooner.

The effect of increased grazing pressure was that the tall tussock grasses (spear and kangaroo grass) were replaced by the low-growing Indian couch grass that appeared to better tolerate the higher grazing pressure. There was insufficient fuel to carry a fire, so pastoralists could no longer use fire to encourage new grass growth. As a result there has been an increase in woody shrubs (currant bush and rubber vine). Extensive sheet and gully erosion now occurs, with bare and relatively impermeable areas and bands of deposited sediment existing adjacent to patches of woody shrubs.

Examples of vegetation decline are also occurring in the higher-rainfall regions of south-eastern Australia. For example, in a number of areas of Victoria (such as the Ballarat district) there has been an increase in the incidence

FIGURE 10.3 Area of bare, eroded landscape in northern Queensland resulting from overgrazing

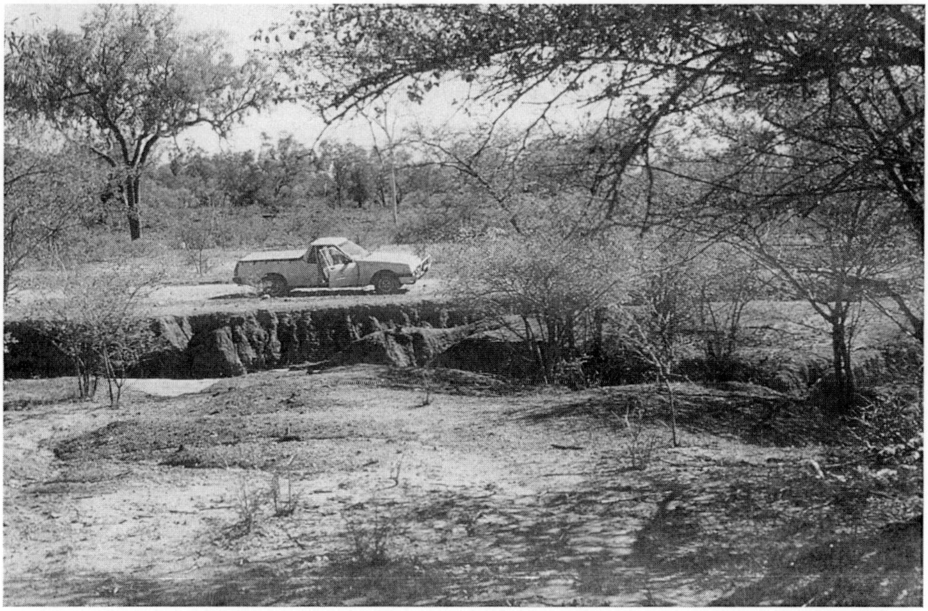

of bent grass (*Agrostis capillarus*) in relatively unproductive pastures on low-fertility soils. This perennial grass spreads rapidly with wet summers and displaces resident pasture species. After flowering in early summer the bent grass produces a stand of extremely low-quality dry feed that sheep and cattle cannot eat. Pasture productivity declines markedly and quickly as stocking rates fall by fifty per cent as a result. Fortunately, a low-cost spray-topping application, using low rates of glyphosate in November, prevents the bent grass from going to seed and overcomes the low-quality summer-feed problem.

Another example of vegetation decline occurs in high-rainfall improved pastures in southern Australia. When pasture improvement occurred in the 1940s and 1950s, there was vigorous, abundant growth of introduced subterranean and white clover when the P status of the soils was increased with superphosphate, and the clovers had a competitive advantage over other species because they could fix their own N and grow in N-deficient soil.

During the next thirty years the legume component in these pastures declined, reducing the productive capacity of pastures. The reasons for the decline of legumes are numerous, complex and interrelated. Periods of low commodity prices and dry seasons reduced income of producers and restricted fertiliser investment. Regular droughts have reduced stocking rates. The variable climate with early or late autumn breaks, and wet or dry finishes to the growing seasons, means that a generally fixed number of stock produce a range of grazing pressures—from extreme overgrazing in a late-break/drought situation to extreme understocking in early-break wet winters and springs. Superimposed on these climatic and 'grazing' stresses have been outbreaks of grasshoppers and earthmite pests, root and leaf diseases, and the continuing competitive pressure provided by well-adapted and competitive weed species. The combination of the stresses has led to legume decline. It is only when combinations of management practices to overcome these stresses are implemented that vigorous, abundant clover growth returns in permanent pastures.

Native grasslands and the biodiversity they foster are resources that play a dual role, helping to meet both private and public objectives. In the past twenty years considerable progress has been made in understanding how to manage native grasses to ensure their survival and at the same time obtain a valuable contribution of feed for livestock, particularly at times when other sources of pasture feed may not be so available. Combinations of intermittent grazing and intermittent light fertilising have been developed that enable native grasslands to help meet both private and public goals.

Pest and disease outbreaks

There have been numerous instances where the productive capacity of farming systems has been diminished by the outbreak of insect pests or plant diseases. A striking example was the arrival of two types of aphids, the spotted alfalfa aphid (SAA) and the blue green aphid (BGA) in 1977. These aphids had a devastating impact on legumes in the genus *Medicago*, which comprises the annual medics and lucerne (*Medicago sativa*). The annual medics are used in pasture leys on alkaline soils in the Mediterranean cropping zone while lucerne is a perennial legume that is widely grown in pastures and as a hay crop. The SAA and BGA aphids decimated medic stands across the country, and the existing medic genotypes could not tolerate the aphids. A huge breeding effort was undertaken to develop new annual medic and lucerne cultivars with enhanced aphid tolerance. Quarantine restrictions were relaxed and many lucerne genotypes were imported from North America, where these aphids were widespread. The plant breeders succeeded and now there is a wide range of aphid-resistant cultivars available to Australian farmers.

Environmental damage

The majority of the population of Australia live in coastal cities. Among this urban population there is a growing environmental awareness and a concern about the well-being of the environment. If practices are damaging to the wider environment, then their continued use will be questioned. A farming system that depends on practices that damage the wider environment is therefore not sustainable. There are a number of environmental concerns associated with current existing farming practices.

Water quality

Productive farming systems in the higher-rainfall regions generally require the use of additional plant nutrients, applied as fertiliser, to maintain adequate plant growth rates. Some 'leakage' of nutrients occurs from the soil–plant or soil–plant–animal system. Nutrients move off the farm in rainfall run-off or drainage water and make their way into surface or underground water bodies. The nutrients that are of most concern are phosphorus (P) and nitrogen (N), which will cause eutrophication of surface water bodies. This results in the increased growth of algae in the water, which can be very unsightly and poisonous. The general perception is that increased P levels in these water bodies are responsible, although the dry summers, lack of water movement and lack of turbidity are mainly responsible.

The problem of N leakage occurs when the nitrate form is leached from the soil into underground water. A build-up in nitrate concentrations in the water can become toxic for humans should that water source be used for human consumption. Such leakage is an inefficient use of N. It contributes to soil acidification, and it can have these adverse effects on the environment. Management practices aim to minimise nitrate leakage.

Pesticide or heavy metal contamination

There are examples where land that had been used for intensive crop or horticultural production has become contaminated by earlier applications of pesticide. For example, there were repeated applications of the highly persistent chlorinated hydrocarbon insecticide dieldrin used in growing potatoes in Victoria and sugar cane in northern NSW in the 1960s and 1970s. Consumers are increasingly concerned about pesticide residues in food. Researchers and farmers develop farm practices that are able to increase productivity and at the same time require less chemical pesticide.

Another concern relates to the potential contamination of soil where high rates of cadmium (Cd) in P fertilisers have been applied in the past. Cadmium is a heavy metal that accumulates in the soil, and can be taken up in different plants to differing degrees. It is present in phosphate rocks and therefore becomes a contaminant of single superphosphate. Lack of awareness of Cd concentrations in P fertiliser meant that high Cd rocks were used in making superphosphate. Again there are lessons here to be learned from the past about potential production problems today. The result in this case has been the move towards using low-Cd rocks for making P fertilisers.

Decline in remnant vegetation

Apart from feeling that the 'world would be a lesser place if we lost even one species', there is the prospect that a particular species might, at some stage, be able to serve a useful purpose for humanity. There are concerns that farming practices are resulting, or might result, in the loss of species. If this is so and a particular species is under threat of extinction, then protective measures need to be undertaken to preserve it. In addition, if there is a small community of remnant vegetation that is considered to be rare and therefore precious, then the public strives to protect and sustain that community. Some native plant communities are useful for farm production, while others are unable to contribute much to farm production.

The native grass *Agrostis adamsonii*, which is adapted to wet saline soils, was thought to be extinct in southern Australia. In 1990 it was found growing in a saline discharge area in western Victoria, and then subsequently located in two other saline sites in the district. The farmers were quite prepared to assist in fencing off the areas to ensure that its continued survival was assured. It so happens that this species appears to tolerate grazing and so it has potential in revegetating exposed saline discharge areas on farms in the region.

Towards more sustainable farming systems

There is much evidence that processes that reduce the quality and productive capacity of land and water are occurring in many paddocks across Australian farmlands. In the previous sections the loss of topsoil from soil erosion, the decline in organic matter in cropping soils with an associated deterioration in soil physical properties, and the depletion of mineral nutrients in crop and pasture soils have all been canvassed. Some soils have lost productive capacity from acidification and salinisation, while many rangeland and improved pastures have experienced a loss of desired species and invasion by weeds, and also become less productive. At the same time, Australian agriculture is producing more than ever before. The land is vastly more productive than when it was first settled by Europeans.

One message frequently put forward is that the high rates of inputs such as fertiliser and pesticides ought to be reduced because at these levels they damage the environment. An alternative low input system is advocated. The fundamental question here is about the impact that reduced inputs will have on productivity.

The first principle of sustainable agriculture is that farm productivity has to be maintained or enhanced. The economic reality is that productivity must be increased in the future because of the need to feed the growing world population, to be competitive enough to earn export income, and to cope with declining terms of trade.

Higher levels of productivity generally reduce the unit cost of production, which makes the farm products more competitive on world markets. The key question from the farmer's viewpoint is whether the value of lost production with low inputs instead of high inputs exceeds the savings from reduced input costs. Certainly for the intensive high-rainfall pastoral and cropping enterprises on the majority of farms, reduced inputs would lead to a fall in output, a rise in per unit cost, and lower profits, which jeopardises the viability of these farm businesses.

Changing conventional farming practices

One objective of farming practices in Australia needs to be that the most serious degrading processes described above are prevented and severe damage corrected, if agriculture is to be truly sustainable. Alternative 'correct' practices are available to achieve these goals. The objectives of an evolving, continually 'improving' conventional agriculture are now to maintain plant cover on the soil surface, minimise tillage to preserve soil organic matter and maintain a favourable soil structure in cropping soil, replace mineral nutrients to offset losses, avoid non-degradable pesticide, avoid excessive grazing of rangeland communities during drought, and replace perennial, high water-use plants to minimise groundwater recharge.

Good use can be made of indicators in agriculture to ensure that sustainability-productivity is 'on course'. Two useful concepts can be borrowed from the discipline of ecology to assist in this area. The first is the concept of 'status'. This is a measure of the existing sustainability-productivity levels. The second is the concept of 'trend'. This introduces a time perspective; it indicates the direction that sustainability-productivity is heading: whether it is declining, holding steady, or improving.

Sustainability indicators of status and trend need to be relevant, simple, accepted and able to be readily interpreted. A range of indicators have been proposed for different agro-ecological regions, for the sustainability of the soil resource, for the environmental impact on water quality, and for chemical contamination of the environment at large. These are all useful. Farm profitability is also a good indicator of how well a business is performing.

Characteristics of the farm system that are important from a sustainability viewpoint can be measured and systems 'scored' for practices involving nitrogen management, amount of tillage, and for stubble retention. A final score gives some measure of the 'well-being' of the cropping system for that paddock. For example, maximum points are for a clean vigorous legume pasture phase (5 points), for total retention of crop residues (4 points), and for direct drilling with narrow points (1 point). A paddock score of 10 indicates highly sustainable management settings for that paddock (although attention would need to be given to acidification possibilities in the medium term). An alternative paddock where cereals were rotated with oilseed crops (0 points), which received a long, cultivated fallow with a disc implement (0.5 points) with all residues being removed by burning (0 points) gives a score of less than 1. This score indicates a non-sustainable management system where soil structure and nitrogen status are declining and the soil is vulnerable to erosion.

Contrasting economic and other views on sustainable resource use

Often, there are differences in the way agricultural scientists and agricultural economists think about sustainability. A scientist might consider that if there is any evidence of a decline in the physical attributes of a resource that is likely to reduce the productive capacity of that resource, then the production system is not sustainable. An economic way of thinking about using resources focuses on the expected benefits versus the costs of alternative uses of such resources, from the viewpoint of the community as a whole. This approach emphasises the roles of all the inputs to production, and the scope for substituting one input for another.

The economic view differs from the more common science view that physical and biological measures of resource degradation directly relate to the welfare of people, and so minimising degradation will automatically somehow produce benefits greater than costs. The economic view is that the physical existence of resource degradation is not necessarily evidence that an economic or social problem exists or will exist in the future.

Production involves many inputs. For instance, the contribution of the characteristics of the soil to agricultural production has been declining in importance for most of the last one hundred years. The components of the soil are just some of these inputs. If nutrients and moisture are not able to be supplied to plants from the soil, they are supplied by the farmer.

Virtually all of the increased production of agricultural commodities in Australia comes not from increased amounts of land producing more output, but from more inputs being added to existing land to enable more output to be produced from that land. The contribution of the soil to production has become relatively less important as other inputs are added, such as artificial fertilisers, herbicides, nematicides, water, improved plant and animal genotypes, and changes in management practices such as reduced tillage or changed crop rotations or changed timing of lambing and so on.

It is common for people to see land (or more precisely, soil) as a fundamental input to production that needs protecting at all costs. Soil is lost, somehow. Resources are changed from what they are now to something different. This is all very 'bad' and the consequences are 'dire'. This view seems to be that the geological and ecological state of affairs ought to be preserved, at least as from now or even returned to what it was at some arbitrary time in the past. But the natural world changes with or without people. So, what is the objective?

The alternative, economic, view is that land is but one of many inputs to production; land, flora, fauna are not pristine-static; it is only by using resources in the ways most highly valued by society that the wherewithal will be found to preserve the environmental features that are highly valued (economic and ecological viability can be complementary, or at least are not necessarily competitive). The economic view is that resource quality and land in particular is seen in terms of their contribution to society's welfare. By this standard, most of Australia's natural resources are in better shape than ever before.

In economic analyses the viewpoint is that of the public interest, with the public defined to include the future generations. Does the current generation owe it to a future generation to bequeath to them a particular paddock with a particular status in terms of soil particle size and soil structure, mineral nutrients, organic matter, vegetative cover and so on? Or does this generation merely have the obligation to pass on to the next generation, the capacity to produce from that paddock whatever is needed and to be able to do so profitably?

Some resource degradation?

An idea that is not obvious is that there is an optimal level of resource degradation. This means that it is rarely economic to try to have no degradation of resources (however defined). Rational farm managers will not spend money on improving the land or reducing degradation unless expected benefits exceed the costs. Degradation of a natural resource is an economic problem if the losses caused by the resource degrading are greater than the benefits that could be produced if the problem was 'fixed'.

The question of running out of resources

Are we going to run out of the resources currently needed for production before they are replaced by superior resources and technologies? Many people think so. Economists generally do not think so.

Natural resources are a basic part of a nation's capacity to achieve economic growth by producing goods and services and expanding the annual gross national product. The quantity and quality of natural resources are relatively fixed compared with the quantity and quality of a nation's labour and capital resources, including knowledge and technology. However, the natural resources of a nation are not absolutely fixed because the available supply of natural resources can be increased by discovery of new reserves of resources. Or, new technology can make it possible to utilise supplies of a resource that could not previously be used. Or, arable land whose productive capacity has been depleted can be renewed.

The price of the resource has a bearing on its rate of use, and its rate of discovery. Generally as a sought-after resource becomes scarce, its price increases. This will reduce its rate of use because it will cost more to use it. At the same time the higher price will increase the rewards for discovering new sources of the resource, which invariably will lead to new discoveries.

'Doomsday prophets' extrapolate growth in demand for resources, and conclude that there will not be enough resources around in the future for people to maintain or improve their standard of living. A major weakness of this view is that it does not allow for linkages between demand, supply and price, where a rise in price caused by a shortage of something causes it to be conserved and created in the following ways:

- Firms and households demand less of a resource whose price rises and less of products using the resource as an input.
- Previously uneconomic resources become economic.
- Exploration to find new reserves of the resource is encouraged.
- Production of known substitute resources is encouraged.
- Research and development to improve known substitutes is stimulated.
- The search for new substitutes is encouraged.

REVIEW QUESTIONS

1. Productivity, profitability and sustainability in agriculture are directly related. Discuss.
2. Explain some indicators of sustainability of resource use.
3. Explain the processes of acidification, dry land salinity, and soil erosion.
4. How does irrigation cause salinity?
5. What problems in relation to nutrient run-off and nutrient depletion are found in Australia's agriculture?

FURTHER READING

Barr, N. and Cary, J., *The Australian Search for Sustainable Land Use: Greening a Brown Land*, Macmillan, Melbourne, 1992.

Cocks, K. D., *Use With Care*, University of NSW Press, Sydney, 1992.

Gardner, C. J., McIvor, J. A. and Williams, J., 'Dry tropical rangelands: Solving one problem and creating another', in *Australian Ecosystems: 200 Years of Utilization, Degradation and Reconstruction*, Proceedings of the Ecological Society of Australia, vol. 16, 1988.

Green, R., Harris, S. and Throsby, D., *Ecologically Sustainable Development Working Group—Agriculture*, Australian Government Publishing Service, Canberra, 1991.

Hamblin, A. P., Environmental indicators for sustainable agriculture, Report on a national workshop, November 1991, Bureau of Rural Resources, Land and Water Resources Research and Development Corporation, Grains Research and Development Corporation, Canberra, 1992.

Parsons, W. T. and Cuthbertson, E. G., *Noxious Weeds of Australia*, Inkata Press, Melbourne, 1992.

Poincelot, R. P., *Towards a More Sustainable Agriculture*, AVI Publishing Co., Westport, Connecticut, 1987.

Russell, J. H. and Isbell, R. F. (eds), *Australian Soils: The Human Impact*, University of Queensland Press, St Lucia, 1986.

WEBSITES

Australasian Agribusiness Review and Australasian Agribusiness Perspectives: www.agrifood.info/

Australasian Farm Business Management Network: www.csu.edu.au/faculty/science/saws/afbmnetwork/

Australian Agricultural and Resource Economics Society: www.aares.info/

Australian Bureau of Agricultural and Resource Economics: www.abareconomics.com/

Australian Farm Institute: www.farminstitute.org.au/

CSIRO: www.csiro.au/

National Farmers Federation: www.nff.org.au/

NSW Department of Primary Industry: www.dpi.nsw.gov.au/

Vic Department of Primary Industry: www.dpi.vic.gov.au/dpi/index.htm

11 Improving Cropping Activities

CONTENTS

Introduction

Continual improvements in crop and animal productivity have to continue to occur on Australian farms in the future. Improving the efficiency of production processes means more output is produced for the same cost, or the same amount is produced but the cost of production is reduced. There are two particularly compelling reasons why this must happen. The first is that the tendency over time is for the prices farmers receive for their production to decline and the costs of production to rise. This occurs as the rest of the economy grows and is bound to continue, albeit with fluctuations around this trend. This means that farmers' continuing viability will depend, as it has in the past, on improving the efficiency of production processes.

The second reason that improvements in productivity are necessary is that Australian farmers must compete with other countries that export primary products in the world economy, and these competitors are continually increasing their efficiency.

There are some perennial dilemmas for agricultural production that relate to changes that accompany every productivity-increasing innovation adopted. First, there is the 'law of the minimum'. As the effect of one part of the production system that is restricting output is ameliorated, other factors become limiting. For instance, there is little point liming soil to reduce acidity if the more serious limits to greater output are deficiencies of phosphorus, or sulfur or nitrogen or minor trace elements such as molybdenum or zinc.

Second, for every step forward in production there is usually some offsetting effect reducing the full extent of the gains from the innovation. For example, increases in production from animals due to higher stocking rates involve higher risk costs from drought, or higher disease costs, or lower prices received because output is greater. In another example, soil pH in parts of southern Australia has declined under permanent pasture where superphosphate has been applied regularly and annual legume plants have been established. The decline in pH can be one pH unit over fifty years of improved pasture. With increasing pH, aluminium and manganese toxicity affect pasture production.

Applying large amounts of nitrogen in intensive horticulture and irrigation cropping increases acidification. Adding lime to ameliorate the problems caused by acidity can enable the next output-reducing problem to come into play. The effects of root disease might then become the factor that is limiting yield. The way of the natural world seems to be that for every improvement in productivity there is an opposite but hopefully not quite equal reaction that reduces the potential benefits.

Improving the profitability of dryland cropping

When it comes to growing annual field crops, the farmer's aim is generally to make as much profit as possible over time. Each individual crop has to be grown as part of a longer-term cropping program that must maintain the productivity of the farm. That is, farmers have to choose combinations of crops that promise to return good profits in the current year, but also have an eye to what this year's combination of crops will mean for crop profits in the next few years. Usually the key to achieving profitable crop returns is to achieve high crop yields. This is because the harvested grain from the crop is sold by weight, although the price paid will depend on grain quality.

Since the beginnings of the Australian grain industry, in the middle of the nineteenth century, six generations of Australian farmers have strived to produce high-yielding crops and earn a living from their farms. Significant increases in yields of wheat, the country's staple cereal crop, have occurred, although climatic constraints limit the magnitude of productivity gains. An examination of past changes in crop yields and profitability is useful for all who seek to make improvements in future crop productivity. The major focus in this dryland cropping section will be on southern Australia where the majority of the rain falls during the late autumn, winter and early spring. Dryland cropping is also practised on farms in northern NSW, and in southern and central Queensland, where the rainfall is summer dominant. Different approaches to improving productivity in the northern cereal belt, compared to the south, are discussed.

The southern cereal belt

The southern cereal belt includes south-western Western Australia within the 270–700 mm mean annual rainfall zone, the southern part of South Australia (250–500 mm), land to the north of the Great Dividing Range in Victoria (300–550 mm) and the southern slopes and plains of NSW where the annual rainfall is 300–600 mm. Looking back over the last 150 years of crop production, the inescapable conclusion is that Australia's crop farmers have been in a continual struggle to come to grips with growing annual dryland crops in the difficult Australian environment. There has been a steady increase in understanding the limitations to crop performance posed by the climate and the soils, and how to manage these limitations. Innovative farmers, agricultural scientists, and inventors have all made major contributions in this area. The outcome has been a continuing change in farming practices to overcome the problems and limitations. But the years have not been easy for the families who have lived and worked on cropping farms: the cyclic occurrence of poor grain prices and recurring droughts take their toll. These cycles will continue. Future production problems will require more changes in farming practice to maintain and improve crop productivity and profitability.

The first fifty years (1850–1900)

Grain farming in southern Australia began to expand during the 1860s as the railway networks reached cropping country. At the same time the land available for cropping increased as a result of government policy to settle land. The shortage of farm labour, long a characteristic of Australian agriculture, was worsened by the gold discoveries in Victoria. The problem was to some extent reduced by the invention of the field harvester and the stump-jump plough, which meant that the area that one person could farm increased substantially. Wheat was the major crop, but the wheat varieties were largely of English origin, which were not well adapted to the growing conditions in southern Australia. The chief problem was the rapid increase in temperature during the spring. In some years the spring was also dry, so the 'hot-dry finish' to the crop was unfavourable for the slow-maturing European wheat varieties. Nevertheless, the early farmers attempted as best they could to maximise profit by growing crop after crop. This practice of continuous cropping led to nutrient loss, as each tonne of wheat removes around 20 kg nitrogen (N), 3 kg of phosphorus (P), and other nutrients from the soil. As a result, grain yields could not be maintained; yields declined as repeated cropping continued to deplete nutrient supply. By the end of the nineteenth century average wheat yields had declined to almost half of what had been achieved from the virgin cropping land in the earlier years. The severe economic depression of the 1890s, and the eight-year drought, from 1895 to 1903, resulted in many farming families losing the battle to farm profitably and, not for the last time, some had to abandon their dream of farming.

The second fifty years (1900–1950s)

After the turn of the century, three changes in farming practices led to a marked improvement in crop yields. The first change was the increasing use of single superphosphate as a fertiliser to supply P, which alleviated the widespread P deficiencies that had developed with continuous cropping on soils that were naturally low in P. The second change in practice was the long fallow, where a crop was grown every second year, after the soil had been maintained in a cultivated, weed-free state during the previous spring, summer and autumn. The fallow generally resulted in higher soil moisture levels at sowing, and increased N availability from the breakdown of organic matter that occurred during the fallow phase. The third change was the use of the new, early-maturing wheat cultivar called Federation, bred by William Farrer. This wheat cultivar matured early and yield was less constrained by the normally hot-dry finish to the wheat crop. Federation wheat was particularly useful in the drier areas of the cropping zone such as in central-western NSW, northern Victoria and South Australia. As a result of improving the N and P status of the soils and the use of better adapted wheat genotypes, crop yields over the next 45 years or so returned to the levels achieved when crops were first grown in virgin soil.

Still, disturbing events were occurring in the wheatbelt. Repeated cultivation to control weeds during the long fallow phase maintained the soil surface as a fine, dry dust, which was thought to reduce the loss of moisture from the soil. Instead it exposed the soil surface to wind and water erosion. Huge dust storms occurred regularly over south-eastern Australia in the 1930s and the dry 1940s, and a great deal of the surface soil was blown away. Continuing cultivation was depleting organic matter levels in the soil, so there was less total N in the soil and a reduction in the stability of soil particles and the pore spaces between them. The physical fertility of the soil was deteriorating over time. Wheat yields per hectare, which had improved in many areas after the turn of the century, had peaked by the 1920s. There were no signs of any yield improvements in the established grain areas up to the Second World War.

The third fifty years (1950–2000)

A new farming technology, legume ley farming, was being developed. Its adoption required favourable markets for livestock and livestock products. The 'wool boom' in the early 1950s led to its rapid and widespread adoption by farmers. The practice was to establish a legume-based pasture for a number of years (usually two to five years) in cropping paddocks, and then return the paddock to crops for up to three years. A rotation was established for a particular paddock where a sequence of crops was followed by a number of years of pasture, before returning to the cropping sequence again. In the higher-rainfall cropping regions where the soils were acidic, the legume used was subterranean clover (*Trifolium subterranean*), while different medics (*Medicago* spp.) were used on drier more alkaline soils. Unfortunately there were no similar successful legumes available for the cropping areas in northern NSW or in southern Queensland, so the benefits of legume leys were restricted to southern Australia. Remarkable improvements in soil N status and physical fertility resulted from the pasture, and fertility could be exploited during the cropping phase to produce high-yielding crops. These crops, in turn, would lower the fertility. In addition, the legume-dominant swards minimised the opportunity for cereal root diseases, such as take-all and cereal cyst nematode, to persist in the soil and infect cereal crops after the pasture phase.

Legume ley farming was highly successful for over a generation. Then the legume phase began to lose its effectiveness, for a variety of reasons. The performance of legumes depends on a dense stand of legume seedlings establishing in the autumn. This requires a large seed-bank of annual legume seed to be maintained in the surface layers of the cropping paddock. The increasing use of crop herbicides, longer cropping sequences, increasing soil acidity and the lack of P fertiliser reduced the vigour and density of the annual legume in the pasture phase. The failure of the legume resulted in N-deficient, grassy, weed-infested pastures. There was an urgent need to find practices that would take over the beneficial, restorative effects of legume-dominant pastures that improve soil N and soil structure, and control cereal root diseases.

Part of the answer to this problem came in the form of the crop management strategy known as conservation farming. Conservation farming has two aims: to reduce the frequency of tillage in crop production and to retain crop residues on the soil surface to protect the soil from erosion. Reduced tillage was a way to improve the stability of soil particles and soil porosity, resulting in a more favourable soil structure for crop growth. Previously the aim of tillage was to control weeds, but it was now possible to control weeds using herbicides. Crops were now sown directly into uncultivated soil, or sown into soil that had been cultivated once or twice after the autumn break. The crops were sown early, and troublesome weeds like annual ryegrass (*Lolium rigidum*) or wild oats (*Avena fatua*) were able to be controlled after crop emergence by grass-selective herbicides. The N status of the soil was maintained by sowing grain legume crops in rotation with cereals and the use of N fertiliser. Cereal root diseases such as take-all and cereal cyst nematode were controlled by growing 'break crops' such as grain legumes or oilseed rape in rotation with cereals. Grass-free break crops could be achieved using post-emergent grass-selective herbicides. There was a trend towards continuous cropping, and rotations were established where the cropping sequence regularly alternated between grain legumes or canola (or both), and cereals such as wheat, barley or triticale. Continuous cropping, together with direct drilling and stubble retention, became a profitable farming system in the more favourable cropping regions of southern Australia.

This seemingly profitable cropping system soon developed major problems. This has been the pattern for all new cropping systems in this country, just when farmers became reliant on them. In the late 1980s and early 1990s the cause of the problem was the development of herbicide-resistant populations of annual ryegrass (*Lolium rigidum*) in cropping paddocks that had been repeatedly sprayed with the grass-selective herbicides. Methods for controlling these resistant populations using a variety of weed control practices needed to be developed.

Entering the twenty-first century

The number of herbicide-resistant weed populations, particularly of annual ryegrass continued to increase. There were now many farms where the repeated use of selective in-crop herbicides had led to resistant weed populations. The new cropping systems required integrated weed management strategies to manage these populations.

However, Australian grain farmers had a very difficult start to the new millennium because the expected rains did not come. There was a major drought in 2002 followed by further droughts in 2006 and 2007 with below average rainfall in most of the intervening years. Rising global temperatures is a looming change for the Australian grains industry to manage.

The dry times, plus the predictions of higher temperatures and increasing evaporation rates, were signalling the need to develop more water-efficient farming systems. Such cropping systems were beginning to develop with an array of sophisticated and expensive technologies that fell under the banner of *precision agriculture*. Such technologies included GPS-guided steering systems on tractors that could control machinery movements and restrict traffic to permanent tracks across the paddock. This reduces soil compaction and loss of soil quality. The adoption of reduced tillage and indeed zero-tillage also accelerated, together with the retention of standing stubble from the preceding crop.

Changes in relative prices of wool, meat and grains meant land uses changed in the high-rainfall zones of southern Australia. An increasing number of livestock producers were now growing grain crops. The use of raised beds was widespread. These ensured that waterlogging, a constraint that previously limited cropping in the wet years in the high-rainfall areas, would no longer cause crop failure. The run of dry years so far has meant that wet winters, which cause waterlogging, have not occurred.

In addition to the need for increased water efficiency, the concept of 'soil health' began to gain prominence in the first decade of the new century. This was all about enhancing the biological fertility of cropping soils, with increasing inputs of organic matter. Such inputs increase microbial activity in the soil with potential improvements in soil aggregation and soil porosity. Improving the biological activity and diversity above the soil surface was also occurring with the increasing use of integrated pest management practices. While these new approaches began to deliver improved environmental benefits in the cropping system, the situation with herbicide-resistant weeds was problematic.

Advances in the genetic improvement of crop varieties followed the enactment of the Plant Breeders Rights Act in 1994. This allowed plant breeders to collect royalties from the new varieties that they had bred. This encouraged private investment, and a range of commercial entities and joint ventures involving private, public and international collaborations began to develop. Many of the important crop species have been the target for improvement with increased commercial access to intellectual property and valuable germplasm. The lifting of the moratoria on the growing of genetically modified crops in Victoria and New South Wales in 2008 has opened the door to a whole new order of genetic manipulation of crop genomes and potential crop performance.

A student of history will note that all farming systems developed so far to produce higher grain yields on Australian cropping land have not been sustained without subsequent changes. Agricultural scientists in the future need to be aware of this pattern, which has continued over the last 160 years; it is likely that the pattern will continue and there will be a continuing need to devise new farming practices to overcome biological, economic and social problems that occur on Australian cropping farms.

Improving sustainable crop yields

The management–yield–profit relationship

The impact of crop yield on farm profits is profound. The reality today is that there are some well-managed farms where high grain yields and good profits from crop rotations are achieved consistently, and there are many farms that are not managed as well and are low-yielding and less profitable. A higher yield of wheat directly increases total gross margin (gross income minus variable costs). For a given level of overhead costs, extra gross margin adds directly to increases in operating profit. The benefits of improving the management of a cropping program are obvious and real, and the way a crop is grown goes a long way towards determining success or failure of a cropping business.

The close relationship between management, yield and profit can be seen in the following hypothetical comparison. Take the good manager who grows a high yielding 4 t/ha wheat crop. He uses new disease-resistant varieties, applies adequate N fertiliser to suit the season, controls all weeds, and sows the crop quickly in mid May. The variable costs of $300/ha and the net price of $230/t, results in a gross margin of $690/ha. In contrast, a poor manager, on similar land next door, grows a 2 t/ha crop due to poor weed control, a late sowing in early July, and low N inputs. Lower variable costs of $200/ha, and a lower price of $200/t, gives a gross margin of $200/ha. Thus superior management led to a twofold increase in yield, and a threefold increase in the activity gross margin for the crop, which go towards covering overhead costs and leaving a good profit.

In the quest for high yields the farm manager must manage the cropping program to reduce the influence of the factors that limit crop yields. The major limiting factors, discussed below, are responsible for reduced grain yields on Australian cropping farms, where average yields are less than 50 per cent of their potential theoretical yields. The critical point is that *all* of these factors must be managed well, if high crop yields, and therefore high rotation gross margins, are to be achieved and sustained into the future. Good managers get everything right.

Using rainfall as efficiently as possible

The first approach to making best use of available rainfall is to grow vigorous, healthy crops in fertile soils. Under these conditions crop yield is related directly to crop water use. This is because water loss from crop leaves by transpiration occurs when leaf photosynthesis (and dry-matter yield accumulation) occurs. This happens when leaf stomata are open. Open stomata enable water vapour to diffuse out of the leaves, and for carbon dioxide to diffuse into the leaves for photosynthesis. This unavoidable loss of water that occurs at the same time as the gain in carbon dioxide for photosynthesis is the ultimate compromise that plants have to make in order to survive on land.

Given the loss of valuable water, which is a consequence of plant growth, the aim is to produce as much grain as possible, per unit of water that is transpired by the crop. The term used to describe the mass of grain produced per millimetre of water lost is *water use efficiency* (WUE):

Water use efficiency (WUE) = Grain yield (kg/ha) ÷ Total water transpired (used) by the crop

The maximum WUE value for wheat over the life of the crop, when there are no limiting factors constraining crop yield, is around 20 kg/mm. This means that for every millimetre of soil water transpired a wheat crop should produce an additional 20 kg of grain per hectare. This figure is not commonly achieved because factors such as weed competition, diseases, pests, or nutrient deficiency, are likely to be constraining yield. The task then for the grain farmer is to manage the cropping system so that these limiting factors do not operate to reduce WUE values for the crop. This is the essence of good crop management.

The second approach to making best use of rainfall is to ensure that the crop uses as much rainfall as possible. This means that much of the rain should end up in the crop's root zone in the soil, where it will be available for the crop to transpire during its vegetative and reproductive growth. This will be significant in the future with climate change. The approach has the aim of 'using every drop (of rainfall) for the crop (for growth and grain production)', and gives rise to the concept of *rainfall use efficiency*:

Rainfall use efficiency = Rainfall transpired by the crop ÷ Total rainfall

In order to achieve high rainfall use efficiency, the grain farmer must manage the cropping system to capture, store (in the root zone), and use as much of the rainfall as possible. This means that water losses from direct evaporation from the soil surface, from rainfall run-off from the soil surface or from lateral flow below the surface, from deep drainage down below the root zone, and from weed transpiration—must all be minimised. Thus rainfall use efficiency involves (1) managing the soil surface to enable it to receive maximum rain infiltration, and (2) managing the soil matrix to enable it to conduct the infiltrating water into deeper soil layers and (3) to increase the soil's capacity to store this water in the root zone for later uptake by crop roots. This objective means that the plant available water capacity (PAWC) of the cropping soil is increased.

There is also the need to increase the capture and storage of rain that falls outside the growing season. If this rain can be stored in the deeper layers in the root zone, it can be used efficiently late in crop growth during grain-filling. The likelihood of storing this out-of-season rain increases if weed growth that follows the rain is controlled when the weeds are small and actively growing.

A range of management practices that focus on improving soil porosity and aggregation will increase the efficiency of rainfall use. The increased pore space enables water to enter the soil matrix and be stored within it, and makes it easier for crop roots to grow through the rooting zone and extract the water and nutrients. Other practices, such as stubble retention on the soil surface, increase rainfall infiltration. These practices are discussed in the section on soil structure below.

Increasing the PAWC of a cropping soil will increase the size of the soil water pool available for crop use. The prospect for increasing the storage of deep water in the soil profile, that the crop roots gain access to after the crop flowers, is also a very attractive proposition, as this water is used very efficiently: the

WUE values for deep water can be up to 60 kg grain/mm. Recent research has found that the placement of organic amendments at 30–40 cm depth in dense clay subsoil resulted in remarkably large grain yield responses, and improvements in subsoil porosity and PAWC. It remains to be seen whether such technology can be commercialised and be able to deliver cost-effective improvements in PAWC for cropping soils with dense clay subsoil.

Responding to rainfall variability

The series of droughts in 2002, and then in 2006 and 2007, highlighted the need to develop flexible and resilient cropping systems. This need will further increase with global warming, which will lead to increased climatic variability. Flexible and resilient systems will help grain farmers respond to variable rainfall. The procedure involves preparing plans that list actions that should be followed in response to different rainfall events. The key objective is to reduce financial losses during very dry seasons and to exploit good seasons and maximise profits when they occur.

Resilient and flexible cropping strategies will be specific to different cropping systems in different regions. Here we will provide examples of different approaches that deal with rainfall variability. The first example is the use of N fertiliser with wheat crops, as this input is generally the most expensive input for the crop. In the past the fertiliser was applied prior to sowing the crop, with no knowledge of how the season would develop. The new flexible approach is to delay fertiliser application until late winter or early spring, when stems begin to grow upwards, and N demand is very high. At this stage of the crop the farmer has a much better idea of how the season is developing, and whether there is any accumulated water in the root zone for reproductive growth. A vigorous crop growing in soil with plenty of available water suggests that high yields are possible. The farmer now has the confidence to apply more N fertiliser to achieve high yields.

Another example of flexible decision making is to delay the choice of crop until the sowing rains actually fall. Consider the situation where the plan is to grow canola in a particular paddock. If there is no sowing rain by the first week in May, then the choice might be to sow a late-maturing wheat variety; continuing lack of rain by early June means that a medium- or early-maturing wheat variety, or perhaps barley, should be sown. Finally the crop choice may be restricted to field peas if sowing rains do not arrive before the end of June. This plan requires seed from different crop types to be available and used depending on when the sowing rains occur.

Another approach might be to sow a portion of the crop in dry soil at the correct sowing time if rains are late. This will mean that less time will be needed to sow the remaining crop when rain does fall, and the length of the growing season for the dry-sown crop will be maximised. Dry sowing works well in very dry, light-textured soil with low weed burdens, when large-seeded crops can be sown deep in the dry soil, particularly if there is subsoil moisture below the crop. However, the strategy can be risky as crop seedlings may perish if small rainfall events occur.

Improving soil structure

Improving the physical fertility of the soil means that its soil structure is more favourable for root growth. This occurs when the soil particles are bound together into different sized aggregates to create a network of pores through the soil matrix. This porous nature makes the soil 'softer' and enables roots to grow easily through the soil. It also helps water and air move through the pore space in the soil to be available for crop roots. Such a soil, with a favourable structure, will be receptive to the rapid infiltration of rainwater and the pore network will create the space to store the rainwater. The soil will be able to receive and store rainfall and will also be good for root growth.

Farming practices have a marked impact on the structure of cropping soils. Excessive cultivation, for example, increases the breakdown of soil organic matter. This fraction of the soil contains the remains of plants and soil organisms. In addition there are the fungal hyphae, bacterial gums, and plant mucilages, which make up the binding agents that 'glue' the soil aggregates together.

Repeated cultivation reduces these cementing substances in the soil, and thereby lower its aggregation and porosity, making the soil more dense. This reduces the aeration under wet conditions and makes the soil harder to penetrate under dry conditions. Grain yields will eventually decline in such soils, particularly those that are sensitive to excessive cultivation, such as the red earth and red brown earths.

It is not surprising, therefore, that when farmers reduce soil cultivation, they find that soil structure becomes more favourable for root growth, and crop yields often improve. This has led the move to minimise tillage in the grain production system. In fact zero-tillage, where the only soil disturbance results from sowing the crop, is being widely adopted across the cereal belt. A consequence of this change is the increased reliance on herbicides to control weeds, as they are no longer being controlled by cultivation.

Other practices, in addition to reduced tillage, can lead to an increase in soil organic matter and a more favourable soil structure. Retaining crop stubbles on the soil surface instead of the earlier practice of burning them, will improve soil structure over time. The stubbles provide substrates for microbial activity that will release cementing substances and increase aggregate formation and stability.

FIGURE 11.1 A hard-setting surface soil in northern NSW

Surface stubbles also increase rainfall infiltration, as they absorb the energy of the falling raindrops, preventing soil particles from being dislodged and soil pores from being blocked. An ideal stubble surface exists when part of the stubble from the previous crop is left standing in rows, and the remaining stubble that passes through the combine harvester is evenly spread across the rows.

New satellite-guided steering technology that can be fitted to the tractor enables the next crop to be sown between the rows of standing stubble so the stubble remains upright until it slowly decays over time.

Another practice to increase soil organic matter is to include a pasture phase in the cropping rotation. Indeed a pasture phase is better able to increase soil organic matter than continuous cropping under zero-tillage.

A third farming approach to make soil structure more favourable for crops is to minimise soil compaction by farm machinery. This involves practices that constitute controlled traffic farming, where the wheels of cropping machinery are restricted to permanent tracks across the paddock.

Satellite-guided tractor steering using GPS technology assists in this requirement. Soil compaction by wheeled traffic reduces porosity; it then absorbs less rainfall and so is more likely to produce rainfall run-off. The reduced porosity means that there is less air space and so the soil is more likely to become waterlogged in wet periods. Similarly the compacted soil is less likely to store water that is available to crop plants.

Farmers who have switched to controlled traffic talk of many observed benefits: these include softer soils away from the wheel tracks, less rainfall run-off and increased infiltration, a shorter time before machinery can be driven along the tracks after rain, less energy and fuel costs to sow the paddock, and most importantly an increase in the WUE values up to and exceeding 25 kg wheat grain/mm of water use. There are many benefits from reducing soil compaction caused by machinery traffic across the paddock.

There are also a number of problem soils in the Australian cereal belt that have inherent soil structural 'defects'. These soils require special management if high grain yields are to be obtained. The first of the problem soils are the sodic soils, in which sodium ions occupy more than 6 per cent of the cation exchange sites in the soil. The sodium ions cause the soil clay particles to disperse on wetting. The dispersed clay particles form crusts on the soil surface on drying. They also block soil pores and reduce water and air permeability within the soil profile. Applying gypsum ($CaSO_4$) to the soil at high rates (1 to 2 t/ha) enables calcium ions to replace the sodium. This prevents dispersion.

The second of the problem soils where unfavourable soil structure limits crop yields are the duplex soils, which occupy large areas of the Australian cereal belt. Duplex soils have a dense clay subsoil, underlying lighter-textured surface soil. Often the clay subsoils are sodic with high sodium levels, while others are quite saline with high concentrations of salt in the subsoil. The physical and chemical constraints impede root growth and this reduces the potential grain yield. The approach to managing these soils is to increase the growth of roots into the clay subsoil. Recent research has shown how the incorporation of organic amendments into the subsoil can markedly increase root growth and crop yields. The approach offers promise for improving crop performance in these soils.

Choosing the right crop in the right sequence

The main consideration in choosing which crop to grow is whether it will make a profit. This means that the crop needs to yield well relative to the costs of growing it, and that the price for the grain needs to be satisfactory. Then there needs to be a satisfactory gross margin, after deducting the costs of growing the crop.

A profitable choice in the southern wheatbelt is to rotate wheat, malting barley and canola. Pulses such as field peas, lupins or chick peas are also

FIGURE 11.2 The surface of a red-brown earth soil following rain. Note the clay dispersion in the ponded water on the surface.

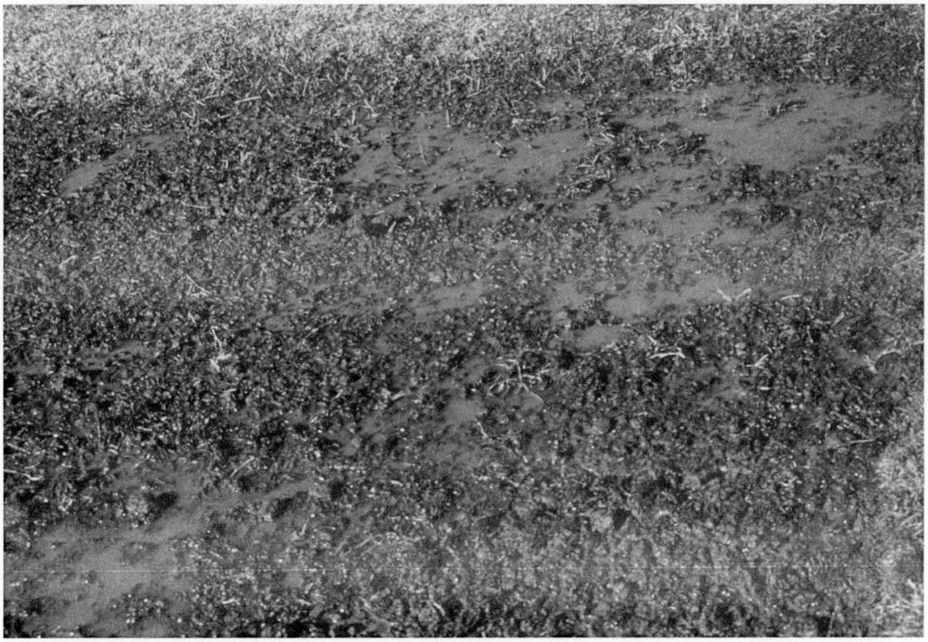

FIGURE 11.3 A similar soil to that shown in Figure 11.2, except that gypsum has been applied to improve the infiltration of water. No surface ponding has occurred after the application.

considerations. While the gross margin from a single crop is important, decision making soon turns to what the crop sequence might be for the paddock. This is all about the rotation of crops that will be followed in the paddock.

Crop rotations are essential requirements in cropping systems. Farmers cannot continue to grow the same crop in the same paddock year after year and expect the yields to be sustained. Crop diseases will soon infect the crop, and yields will decline. There are root diseases such as take-all, caused by the fungus *Gaeumannomyces graminis*, and cereal cyst nematode (CCN), caused by infection of cereal roots by larvae of the nematode *Heterodera avenae*. In addition there are diseases that infect the shoots such as yellow spot (*Phaeosphaeria nodorum*) in wheat, net form of net blotch (*Pyrenophora teres*) in barley, and blackleg (*Leptosphaeria maculans*) in canola.

Growing a different crop species such as a brassica oilseed or a pulse crop following a cereal, provides a break from the cereal disease, as the disease can no longer survive in the paddock. Such crops are therefore called 'break' crops. While new crop varieties are being released with increased resistance towards different diseases, the choice nevertheless is what the appropriate rotation will be to ensure that diseases do not build up in the crop.

In addition to disease considerations, other factors influence the choice of crop sequence to grow in the paddock. Weed control issues arise, where a particular crop needs to be grown to enable a particular weed control strategy to be implemented. An example here is the need to grow a canola crop, so that triazole herbicides can be selectively used to control silver grass (*Vulpia* spp.), which can build up in zero-tilled paddocks in dry years.

Another consideration in choosing a crop is the benefits resulting from improving soil fertility. An example is a grain legume pulse such as peas or lupins, which have the capacity to fix nitrogen in their roots. Some of this fixed nitrogen may become available for the following crop, and thereby reduce the need for nitrogen fertiliser treatments.

On the other hand, a decision to include a hay crop, or to harvest the crop as hay, will have a negative impact on soil fertility. This is because the increased loss of nutrients from the paddock in the hay will need to be replaced to maintain the nutrient status in the soil.

The choice of cropping sequence depends on a whole range of issues. The profit generated by the total sequence over time is of prime importance, bearing in mind that there will be biological benefits of one crop on another within the sequence. Given that yields and associated profits must be sustainable over time, the rotation and associated inputs must deliver benefits in disease control and weed management and must maintain soil fertility over time. Continuous cropping is possible because weeds, pests and diseases are able to be controlled. Careful attention, however, has to be given to maintaining soil fertility.

Managing nitrogen and other nutrients

Nitrogen along with soil water is the key to increasing the grain yield in non-legume crops. Providing extra N to the crop will increase yield in wet years. Crops must have high N supply if they are to produce large yields. A further consequence of high yields is that large amounts of N are removed in the grain. The N concentration in wheat grain is around 2 per cent, so around 20 kg N will be removed in every tonne of harvested grain. This means that a focus on N management is required for high-yielding non-legume crops.

FIGURE 11.4 The effect of nitrogen fertiliser (plot on the right) on the growth of oats

Practices for managing the N supply for a specific crop are established. The approach is to balance N supply with the likely N demand by the crop. For example, the N demand by a wheat crop can be assessed before stem elongation by estimating the crop yield. This can be determined from the available water in the soil from winter rainfall and from an estimate of likely spring rainfall. The N supply side can be determined from a deep soil test, sampled to 60 cm, which measures nitrate N currently available in the soil. The likely additional N that will be supplied from the breakdown of soil organic matter during the crop growth period is estimated. Any shortfall in N supply can be determined by subtracting the estimated N supply from the estimated N demand. This should be applied as N fertiliser during the stem elongation phase of crop growth.

There are a number of ways to build up the N status in cropping soils. Perhaps the most reliable way is to have a vigorous, legume-dominant pasture phase in rotation with grain crops. This option is not used widely as the productivity and profitability of the pasture phases on grain farms can be quite low. In mixed farming, decisions about crops are intricately related to the profitability of complementary activities, such as wool, lamb, or beef production on these pastures.

A second way to increase N supply is to include grain legumes in the cropping rotation. The fixed N_2 added to the soil from a grain legume crop varies, depending on the amount of legume N that is fixed in the root nodules, and the proportion of this N that is removed in the grain at harvest. A further approach is to add N in green manure crops or as organic manures, but this will depend on the availability of such sources.

Other mineral nutrients need managing too. For example, P needs replacing regularly to match P removed in grain harvested. Low P soils, which require additional P, can be diagnosed with soil tests. Potassium (K) requirements can also be determined by soil tests, and will most likely be required on light-textured sandy soils in high-rainfall zones, where there is a history of hay making. Sulfur (S) is required in large amounts by canola, and can be assessed by soil tests. Zinc (Zn) and copper (Cu) are trace elements that can limit crop yields if they are in low supply. Most alkaline soils are low in available Zn, while Cu deficiency is often seen after waterlogging, particularly in wheat grown on acidic duplex soils. The Cu and Zn status in the crop is best determined by tissue testing early in the crop growth phase.

Establishing the crop on time

High yielding crops, particularly crops grown during the winter, have to be sown at the correct time. This is during the second half of May for medium-maturing wheat varieties. Sowing crops too early in April increases the risk of frost damage during the ear-emergence and flowering stages of growth, and grain-set can fail.

On the other hand, crops sown late run the risk of low yields from accelerated development during grain-filling, caused by increasing temperatures and dry conditions in the late spring. The Australian grain farmer faces the difficult choice: sow too early and face increased frost risk, or sow too late and risk excessive heat during the temperature-sensitive grain-filling stage.

Steps can be taken to reduce the risk from frost. These include mapping the areas of the farm that are prone to frost, where the most severe 'frost pockets' occur depending on topography. These areas should be sown late, perhaps with later-maturing wheat varieties, or even barley that can tolerate frost more than wheat. Also it is wise to avoid really high inputs of fertiliser on these high-risk areas. Crops with high N tend to be damaged more by frost than crops with lower N supply. If the crop does become frosted, consideration should be given to baling it for hay to reduce economic loss.

High-yielding crops only occur if the population of crop plants is sufficiently dense. Increasing yield requires increasing density of plants. For example, if the anticipated wheat yield is 1 tonne per hectare then a population of 50 plants per square metre might be sufficient, whereas under the same conditions, a 5 to 6 tonne crop would require a population of around 200 plants per square metre to ensure that yield was not restricted by low plant populations. The required plant population requires an appropriate seeding rate to be used and that seedling emergence is not unduly impaired. The required seeding rate is determined from the equation:

$$\text{Seeding rate (kg/ha)} = \frac{\text{Required plant population (plants/m}^2) \times \text{Grain weight (mg/seed)}}{\text{Expected rate of emergence (\%)} \times 10^{-2}}$$

In this equation the unknown factor is the expected emergence rate. Emergence difficulties occur in soils where the tilth of the seedbed is coarse because of a 'clumpy or blocky' structure of the soil, or where surface crusting develops with post-sowing rain due to dispersive or slaking surface soils.

Poor emergence of seedlings will also occur if the control of sowing depth is poor and seeds are sown too deep. Low emergence rates occur if summer crops that are sensitive to cold temperatures, such as grain sorghum and cotton, are sown too early in the spring and low soil temperatures restrict seedling emergence. Using press wheels to compact soil around the sown seed improves emergence of seedlings in soils where the tilth (or the soil surface aggregate sizes) is coarse.

Controlling weeds

Crop yields decline as the population of weeds in a crop increase. Weeds compete with crops for moisture, nutrients and light. Problem weeds build up rapidly in a crop, compete effectively and reduce its yield, and are difficult to control in the crop. In the case of cereal crops the most troublesome grass weeds in the southern cereal belt are wild oats (*Avena fatua* and *Avena ludoviciana*) and annual ryegrass (*Lolium rigidum*). While small populations of these grass weeds have minimal effect on yield, their population builds up rapidly if they are allowed to set seed in the paddock.

A wide variety of herbicide groups, with different chemical modes of action, have become available to control weeds over the last fifty years. Herbicide manufacturers developed the different herbicides that are selective for weeds but do not affect crop plants. They are generally environmentally safe, and can be long-lasting with extended residual activity if required. The switch to minimum or no-till cropping systems means increasing reliance is placed on herbicides to control weeds. Weeds can no longer be controlled by cultivation, and selective herbicides are used to manage them.

The widespread use of selective herbicides in southern Australia since the late 1980s has resulted in the world's greatest herbicide resistance problem. This has occurred with annual ryegrass, which existed in large numbers across the cereal belt. It had previously been grown widely as a pasture grass on many mixed farms. The change to more cropping meant that 'ryegrass the pasture grass' became 'ryegrass the crop weed'.

Weed control during crop growth using the grass-selective herbicides, the 'fops' and the 'dims', was initially very successful. Their repeated use led to the rapid development of resistance, as the plants with resistance genes were the only ones to survive and set seed. The problem was widespread and it led to a weed control program known as *Integrated Weed Management* (IWM) to manage resistant weed populations and to prevent resistant populations from developing. The approach uses a range of weed control tactics to prevent ryegrass from setting seed. These included cultivation practices, non-selective herbicides, green and brown manuring, hay and silage making, and pasture phases where the ryegrass is spray-topped to stop seed set. Most importantly there is no repeated use of the selective herbicide.

A new approach to weed management being adopted by leading farmers developed from the battles with resistant ryegrass. Farmers learnt that resistant ryegrass could be eliminated if seed-set could be stopped. Two to three years

FIGURE 11.5 Annual ryegrass

of seed-set prevention virtually eliminated the weed population, because seed numbers in the soil seed bank became depleted over this period.

This led a small number of farmers to strive towards 'zero tolerance' of problem weeds. The approach is to develop strategies and tactics in paddock management to stop seed-set and then over time deplete the seed bank, until there are no more viable weed seeds.

An issue here is that some tactics to prevent seed-set in weeds may conflict with other management practices. For example, cutting weedy portions of the crop and making hay or silage may mean that machinery will move off the permanent tracks where controlled traffic is used. Another tactic is to burn chaff from header trails in the autumn to destroy weed seed that has passed through the header; burning stubbles does not fit well with a zero-tillage, stubble-retention system. Compromise or careful planning is required to prevent weed seed-set and preserve soil structure.

Destroying the weed seed bank can be a challenging task. This is the case for weeds like wild radish (*Raphanus raphanistrum*), which has a persistent seed bank. The bank may take six to seven years of seed-set prevention to deplete the seed, by ensuring that absolutely no new seed enters the bank. Then continual vigilance is required as just a few escaping wild radish plants can produce large numbers of seed, and the battle with the weed has to recommence.

Controlling diseases and pests

Crop diseases pose continuing threats to high-yielding crops. The crop plants, growing in dense, uniform populations under warm and moist conditions in the spring, provide an attractive host for the disease pathogens. Not surprisingly, there are an array of pathogens that infect the main crop species, which include wheat, barley, canola and the pulses. Australian grain producers, together with plant pathologists and advisers, are taking the battle right up to the pathogens, using a combination of management strategies.

A three-pronged integrated plan is generally employed for shoot diseases. This involves the use of cultural practices, genetic resistance in the crop plants, and the use of fungicides in the case of fungal pathogens. One example is the stripe rust (*Puccinia striiformis* f.s.p. *tritici*) pathogen, which infects wheat. Farmers are encouraged to control volunteer wheat plants (known as the green bridge) during the summer to reduce the supply of fungal spores that may cause outbreaks of stripe rusts in following wheat crops.

The most resistant wheat varieties should be sown. Early-sown crops can be protected from autumn and winter infections with fungicide-treated seed or fertiliser. Crops need to be monitored closely for signs of infection so early decisions on the use of fungicide sprays can be made before the disease becomes severe in the crop.

A range of root diseases infect cereal crops in southern Australia. Two of these include take-all, caused by the fungus *Gaeumannomyces graminis*, and cereal cyst nematode (CCN), caused by infection of the cereal roots by larvae of the nematode *Heterodera avenae*. Take-all and CCN require a grass host-plant to persist in the soil. Both of these pathogens can be controlled by crop rotation, using non-grass break crops. The third root disease in southern Australia is *Rhizoctonia*, which occurs when crop roots become infected with the fungus *Rhizoctonia solani*. This fungus infects all crops including grain legumes, cereals and canola. This means that crop rotations, with non-grass phases, do not control this disease. The fungal hyphae persist in the soil and infect the seedlings of crop plants after they germinate and emerge. The effects of the disease can be reduced if the soil, together with the network of fungal hyphae, is disturbed by cultivation. Farmers who direct-drill crops modify their combines with narrow points to disturb the soil below the seed placement layer. *Rhizoctonia* damage in crops is also reduced by adding Zn fertiliser to the crop as zinc deficiency appears to make the crop more susceptible to fungal infection and damage.

FIGURE 11.6 The reduction in the growth of wheat plants due to cereal cyst nematode damage to the roots

FIGURE 11.7 Improvement in the growth of wheat in a paddock infested by cereal cyst nematode when an effective nematicide was applied

Insects and other pests also pose threats for grain crops in Australia. The recommended response by the grains industry to these threats has a real environmental focus and involves an approach called *integrated pest management* (IPM). Earlier responses relied very much on the use of broad-spectrum chemical pesticides. Their use led to pesticide resistance by the pests, and to negative community responses to their use.

The basic tenets of the new IPM approach are (1) to correctly identify the pest, and beneficial predator species, which prey upon the pest, and (2) to intervene when the pest population reaches a damaging threshold level using a 'soft' pesticide that only targets the pest organism, and does not affect the beneficial organisms. Experience shows that the population sizes of the beneficial organisms are generally small in the first year of an IPM program. After three to four years they increase to significant numbers so long as the soft pesticides, which do not damage them, are used. The beneficial predator organisms are able to control the pests most of the time, and the threat to crop yields diminishes.

The northern cereal belt

The northern cereal belt includes dryland cropping areas from the north-western slopes and plains of NSW, through southern Queensland including the Darling Downs and lands to the west of the Downs, and up to the central highlands of Queensland. Dryland cropping in the north commenced much later than in the south.

The first crops were grown in the eastern Darling Downs early last century. Grazing lands to the west were converted gradually to cropping. Much of the expansion, particularly on the brigalow lands of southern and central Queensland, occurred in the 1960s and 1970s. The mean annual rainfall in the region varies from 450–1000 mm, with the predominance of a summer rainfall incidence increasing from the south to the north. The major differences in cropping practices and soil requirements that occur between the northern and southern cereal belts result from the differences in rainfall distribution.

Implications of summer rainfall

The rainfall pattern of the northern cropping areas does not suit wheat and other winter-growing crops. This is because most of the rain falls during the summer and wheat grows during the winter and spring when rainfall is more variable in the north. To ensure that there is sufficient moisture for the winter crop it is necessary to accumulate and store additional moisture in the soil. This is achieved by fallowing over the summer and storing as much of the summer rain as possible for the following crop that will be sown in the late autumn or early winter. The key requirement is that the soil has an adequate moisture-storage capacity: deep, heavy-textured clay soils on the Darling Downs and on the northern plains of NSW are ideal for this purpose.

The increased incidence of summer rainfall in the northern cereal belt means that farmers do have the choice of growing either a summer crop such as sorghum or sunflowers, or a winter crop of wheat or alternatives such as barley, canola and chickpeas. Other crops include canary, linseed and triticale in the winter, and mungbean, millet, and panicum in the summer months, indicating that there are a number possible crop options. Long fallows ensure that sufficient soil moisture accumulates when a switch is made from summer to winter cropping. Alternatively, the practice of opportunistic 'double cropping' is used where a second crop is grown immediately after the preceding crop when sufficient rain falls to replenish the soil moisture reserves before the second crop is sown.

Sorghum has now become the major summer crop for the northern wheat-belt and the demand for sorghum grain is likely to remain strong due to its use in beef feedlots and in ethanol production. As with cereal production in the southern cereal belt, the quest for high sustainable sorghum yields is based on the efficient capture, storage and use of rainfall. The yield potential varies from above 6 t/ha in the higher-rainfall, cooler areas where 400 mm might be used by the crop with a water use efficiency (WUE) of 15 kg/ha/mm. In the western areas with higher temperatures and evaporation rates, the potential yield can still exceed 3 t/ha, when 300 mm of water is used with a WUE of 11 kg/ha/mm. Achieving these potential yields requires the sound farming practices outlined for the southern cereal belt. These include using zero-tillage, stubble retention with controlled traffic, together with sound N, weed- and pest-management practices. Sowing the sorghum early, in late September or October, to prevent the crop from flowering in the summer heat, will improve crop yields.

Limited use of pasture leys—implications for soil fertility

The northern cereal belt lacks suitable legume species to grow in a pasture ley in rotation with cropping phases. The unreliable autumn and spring rains and the drier winters, particularly in the north, mean that subterranean clovers and annual medics, which provide effective pasture leys in the south, are not suitable in the north.

The best pasture legume option on deep, well-drained soils is lucerne (*Medicago sativa*). It is not widely used as it can be difficult to establish and many grain farmers prefer to avoid the need for fencing, rotational grazing, watering points and access to a shearing shed in the case of sheep. Establishing small-seeded summer-growing grasses can be difficult because the seed needs to be sown close to the surface, and surface soil dries out rapidly following rain, particularly during the warmer times of the year.

Continuous cropping, interspersed with short and longer fallows, is the pattern of land use on cropping farms in the northern cereal belt. Repeated tillage practices have been the major strategy in the past to control weed growth in short (December to May) and long fallows (May to May with a sorghum–wheat sequence, or December to October when wheat is followed by sorghum).

There is increasing use of minimum or zero-tillage where chemical fallows are maintained with herbicides. The past use of continuous cropping with repeated tillage across the northern cereal belt led to the decline in soil organic matter levels and fertility over time. Declining organic matter and N concentrations have meant that N fertilisers have to be increasingly used to maintain N status. The lack of a suitable legume ley, and the minimal N contribution from grain legume crops, increase reliance in fertiliser N. For example, a high-yielding sorghum crop of 4–5 t/ha will remove 70–80 kg N/ha, so N fertiliser management becomes very important.

The increased erosion threat—the need for stubble retention

High-intensity summer storms in the northern cropping region pose major erosion risks for the farms. Retaining crop residues on the soil surface during short and long fallows is even more important in the north than in the south.

Strip cropping is a strategy to combat soil erosion and is common in parts of northern NSW and the Darling Downs. This involves planting alternate summer and winter crops in narrow strips across gently sloping cropping country (less than 2 per cent slopes). Newly sown, erosion-prone crops are separated by erosion-resistant strips of crop or crop residue. Cropping in the central highlands of Queensland faces increased erosion risks because of the summer rainfall intensity. Increasing emphasis is placed on careful crop stubble management. Crops that produce low stubble loads are grown following crops that produce large amounts of stubble. For example, sunflowers are sown into wheat stubbles so there can be protection for the soil surface.

Weeds and diseases in the north

In general the root diseases in the north are less of a problem than the take-all, CCN and *Rhizoctonia* diseases in the southern cereal belt. Root diseases in wheat such as crown root rot and crown rot do occur and are controlled by suitable rotations. Other diseases include yellow spot in wheat and phytophthera root rot in chickpeas. These are controlled by rotating crops. The threat of foliar fungal diseases, such as stripe and stem rusts in wheat, is greater in the north where temperatures are generally higher in the spring.

Growing crops during the summer means there is a whole suite of different summer-growing weeds to control in northern cropping regions. These include weedy species from the genus *Urochloa*, *Echinochloa*, and *Digitaria*. Wild oats is the major grass weed in winter cereal crops in the north. A major concern

FIGURE 11.8 Effective control of wild oats in wheat using summer cropping in northern NSW (plots on left) compared with plots where continuous wheat was grown (plots on right). The use of pre-emergent or post-emergent herbicides was not as effective as the summer crop rotation.

is the increasing number of wild oat populations that are becoming resistant to herbicides, particularly the 'fop' Group A herbicides. These can be a real problem for winter-growing rotation crops such as chick peas, which compete poorly with wild oats. Again the necessary management approach is to deplete the wild oats seed in the soil seed bank. Such strategies as the use of winter fallows leading into or out of summer crops, together with low-risk knockdown herbicides such as glyphosate and paraquat, are options for an integrated weed management approach.

REVIEW QUESTIONS

1 'One and a half steps forward and one step backward' is one description of the process of progress in agriculture. Is it true?
2 What have been the major steps forward, and backward, in cropping systems in southern Australia over the past 150 years?
3 Explain what is meant by rainfall use efficiency and ways that this measure of efficiency can be improved on cropping farms.
4 How would you compare high- and low-input farm systems?

FURTHER READING

Davidson, B. R. and Davidson, H. F., *Legumes: The Australian Experience*, John Wiley and Sons, Brisbane, 1993.

Loomis, R. S. and Connor, D. J., *Crop Ecology: Productivity and Management in Agricultural Systems*, Cambridge University Press, London, 1992.

Pratley, J. (ed.), *Principles of Field Crop Production* (3rd edn), Oxford University Press, Melbourne, 1994.

Wheeler, T. L., Pearson, G. T. and Robards, G. E. (eds), *Temperate Pastures: Their Production, Use and Management*, Australian Wool Corporation/CSIRO, Australia, 1987.

WEBSITES

Australasian Farm Business Management Network: www.csu.edu.au/faculty/science/saws/afbmnetwork/

Australian Farm Institute: www.farminstitute.org.au/

CSIRO: www.csiro.au/

NSW Department of Primary Industry: www.dpi.nsw.gov.au/

Vic Department of Primary Industry: www.dpi.vic.gov.au/dpi/index.htm

12 Improving Grazing Activities

CONTENTS

Introduction

Australia's extensive animal industries are based largely on grazing of natural and introduced pastures and crops grown for feed. Improving productivity in extensive animal industries requires understanding, changing and managing the animal-pasture complex. This encompasses pasture and animal types and genetic worth, systems of grazing management, and efficiently utilising key inputs such as feed, labour and capital.

Improving profitability from grazing land

The past use of grazing lands in southern Australia

Grazing is the biggest user of land in Australia, spreading across the vast arid inland regions that have low levels of productivity. Mixed populations of pasture plants, which are repeatedly defoliated by grazing animals, and subjected to variable climatic patterns, are dynamic and unstable. Such populations require enlightened management to maintain their productivity and to ensure that they persist over time.

The arrival of European settlers in the high-rainfall regions of southern Australia with their sheep and cattle had a marked effect on the grazing land.

Productivity declined when mistakes in land management were made by landholders as they battled for profitable outcomes. The causes can be attributed to both a lack of understanding of the complex ecology of grazed pastures, and to external circumstances beyond the control of the landholders. The gains in productivity during the twentieth century can be credited to the creativity of the early agricultural scientists and to the inventiveness of Australian farmers.

The decline in native grasslands (1800–1940)

The first Europeans encountered attractive, open grasslands containing dense, deep-rooted perennial grasses. The Indigenous people used fire to produce fresh grass regrowth that would attract marsupial game. Repeated burning of the grasslands suppressed weed and tree growth. Sheep flocks grazed upon these open grasslands. Overgrazing and selective grazing by increasing numbers of sheep caused some grasses to disappear, creating open swards with bare spaces and the beginnings of soil erosion. This process was further exacerbated by fences to control sheep movement.

A litany of further mistakes occurred: rabbits were introduced to provide a hunting experience; they thrived and spread across the land. Pasture weeds were introduced unintentionally in imported fodder (thistles, annual broadleaf weeds and grasses) or intentionally as ornamentals (blackberry, prickly pear, St John's Wort). These also spread and invaded the native grasslands. After the 1944–45 drought the native pastures were degraded seriously: the sparse pastures consisted of shallow-rooted annual species for the most part, which were heavily infested with rabbits, and contained bare and eroded areas.

Pasture improvement in the 1950s and 1960s

A remarkable turnaround in the prosperity and productivity of high-rainfall pastoral farmers occurred during the 1950s and early 1960s. This came from pasture improvement adopted by many farmers, which involved establishing adapted legumes, and in some instances productive perennial grasses, in the pasture paddocks and applying single superphosphate. Subterranean clover was used in the 400–700 mm annual rainfall zone, with white clover being used in wetter areas with more humid summers. The legumes added fixed N to the soil, which greatly increased the growth of grasses. They also improved the quality of the feed from pasture. The perennial grasses extended the growing season and improved the stability of production. The net effect of the so-called 'sub and super' combination was a spectacular three- to fivefold increase in animal production per hectare. The competition from rabbits was virtually eliminated by the release of the *Myxomatosis* virus in 1949. Commodity prices were generally high, particularly for wool, and government incentives to encourage rural exports made all investments in pasture development 100 per cent tax deductible in the first year. Finally the seasons were good, with favourable rainfall patterns during the 1950s and early 1960s.

The decline in improved pastures 1970s–1980s

Serious droughts occurred in eastern and southern Australia in the late 1960s, after a run of good seasons. Feed shortages developed quickly on heavily stocked improved pastures and many improved pasture species were lost and had to be resown after the drought. Low commodity prices were common in the 1970s and the price of single superphosphate increased rapidly. Superphosphate usage declined and then fluctuated about a low base throughout the 1980s and early 1990s. A very severe and widespread drought occurred in 1982, further damaging the pastures. Not surprisingly the density and the vigour of the clover component in temperate pastures declined. The clovers had to compete with a whole suite of annual grasses and other weeds, which seemed to have the competitive advantage in these underfertilised, undergrazed pastures that were deprived of inputs—not so much by farmer choice, but by circumstance. Further pressure was placed on the clovers by a build-up in leaf and root diseases and insects such as red-legged earthmites. There was increasing concern about the poor quality of pastures, about the lack of clover and about the widespread occurrence of N deficiency in pasture soils. Pastoralism during much of the 1970s and 1980s was characterised by a run-down in pasture quality and quantity, accompanied often by a decline in farmers' hopes for their pastures and also in their pasture management skills.

FIGURE 12.1 A degraded pasture in south-western Victoria containing almost no subterranean clover. The pasture had been improved with subterranean clover about 30 years before.

Pasture improvement in the 1990s and into the twenty-first century

As if born out of necessity, changes began to occur during the late 1980s. Scientists at the Pastoral and Veterinary Institute at Hamilton in western Victoria, changed the layout of a grazing trial to ask the question: How many spring-lambing ewes, under varying grazing pressures, can be run on improved pasture when the rate of P fertiliser is increased? It was found that a large number of ewes could be run, over a number of seasons, under high stocking rates, when high rates of P fertiliser were applied. The true potential of combining (1) high levels of soil fertility, (2) high stocking rates of productive livestock grazed judiciously with (3) legumes and perennial grasses that are both productive and responsive to enhanced soil fertility and heavy grazing pressure, began to emerge.

What surprised most observers was the condition of animals grazing these intensive pastures: the fact that the animals were generally in good condition was attributed to the high quality of the heavily grazed, high-fertility, legume-grass pasture. This reflects in part the notion that 'the more that stock eat grass, the more grass will grow, if the soil fertility is up to the task and there is adequate rainfall'. A new 'productivity' package emerged. To the 'sub and super' combination was introduced a third component: appropriate grazing management.

FIGURE 12.2 The grazing experiment that revived the industry—the long-term phosphate trial at the Pastoral and Veterinary Institute at Hamilton, Victoria, in October 1992. The high-P, heavily grazed strip on the left carried 22 ewes per hectare, while the low-P strip on the right could only carry 5 ewes per hectare. Some supplementation was required for both treatments.

FIGURE 12.3 The rate of pasture regrowth by responsive species (subterranean clover, phalaris, perennial ryegrass) in the high-P, heavily grazed pasture in the Hamilton grazing experiment. This occurred over two weeks in October.

FIGURE 12.4 Little regrowth by non-responsive species in the low-P, lightly grazed plot during the same period

A series of grazing trials in Victoria in the late 1990s helped to quantify the benefits from rotational grazing compared to set stocking. The benefits involve better utilisation of pasture, higher stocking rates and enhanced persistence of perennial grasses such as phalaris. Persistence of perennial grasses became a topical issue after the turn of the century with droughts in 2002, 2006 and 2007. Poor persistence has been a problem with ryegrass pastures, whereas phalaris pastures have generally persisted quite well. Many landowners struggled to manage the feed deficits for their stock during the droughts, and then had to face the prospect of resowing paddocks where the pasture failed to persist.

Improving high-rainfall pastures in southern Australia

Stocking rate—the source of profit driver

The key source of profitability from grazing land is stocking rate. Gross value of animal product per hectare, and cost of production per unit of product are key indicators of the economic performance of the grazing enterprise. This is the number of animals per hectare that are grazed in a paddock. Stocking rate determines the amount and value of animal product from the farm per hectare. As stocking rates increase, the quantity of animal product produced increases. This will generally lead to a lower *cost of production* per unit of product.

Grazing animals such as sheep and cattle consume pasture to acquire energy and other nutrients, in order to maintain body function, and to produce product in the form of meat, wool or milk. Stocking rate is standardised across grazing enterprises by converting the numbers of different animal types in dry sheep equivalents (DSEs), which in pasture consumption terms is equivalent to one mature 45 kg Merino wether (castrated male sheep) in average condition.

There is a limit to the numbers of animals that can be run to avoid consuming all the pasture in the paddock. This leads to the concept of *sustainable stocking rate*, achieving a balance between the consumption of as much pasture as possible by as many animals as possible, and the need to leave enough pasture cover in the paddock during dry times to protect the soil surface from erosion.

The issue is how to arrive at an optimum sustainable stocking rate for a particular paddock on a grazing property in southern Australia. The number of stock that can be carried directly relates to the amount of pasture that can be grown, its quality, and when it grows during the year. Many factors affect pasture growth such as soil type and soil fertility, pasture type, and the climate, which includes rainfall and the temperature regimes. The rainfall over which the land manager has no control is the key determinant of pasture growth.

Guidelines for paddock carrying capacity, indicating how many stock (DSE) can be carried per hectare, have been developed from the average annual rainfall for the farm. One fairly conservative approach, which assumes the pasture is well managed and that no growth limitations occur, uses the formula:

Carrying capacity (DSE/ha) = [Annual rainfall (mm) – 250] × 1.0

The formula determines the potential carrying capacity for pasture land with average annual rainfall of 500, 600 and 700 mm to be 10, 14, and 18 DSE/ha respectively. Such determinations are a useful guide to the producer. This approach assumes that all paddocks are the same and does not take into account factors that affect rainfall-use efficiency, such as soil depth and the soil's capacity to store, or topography that affects rainfall run-off. These factors can affect the amount of water stored in the soil and so the length of time that a pasture can continue to grow in the late spring – early summer period when the rains cease to fall as the growing season comes to an end.

A more accurate estimate of a paddock's carrying capacity is based on the length of the growing season. For example, paddocks with growing seasons of 5, 6 and 7 months duration, have estimated carrying capacities of around 11½, 15 and 18 DSE/ha respectively. The number needs to be adjusted downwards if stock are grazed in large paddocks in excess of 20 hectares, as the utilisation of the feed declines because of uneven grazing in larger paddocks.

A useful guide that assists the producer to determine whether the stocking rate matches the paddock carrying capacity is to assess the amount of dry pasture remaining in autumn, which is the time when the pasture cover is most critical in southern Australia. There is little summer rain (and therefore pasture growth) after the late spring. The carry-over dry feed from the previous spring flush is all that can sustain the animals, until pasture begins to grow again following the late autumn rains. Overgrazing in the early autumn, removing much of this pasture cover, exposes the soil surface to erosion. This means that stocking rates are too high.

Stocking pressure under these conditions needs to be reduced to avoid the potential environmental cost of losing topsoil if heavy rain was to fall. On the other hand, an undergrazed pasture results in a large surplus of dry pasture from the previous spring remaining at the end of the autumn, which might still be visible in August–September. Undergrazing in these conditions will result in a loss of quality and quantity of new pasture regrowth, and lost economic opportunity to run more stock and make more profit.

There are currently many low-producing pastoral farms in temperate Australia, and indeed tropical Australia, where there is potential to improve pasture production and increase stocking rate. Such actions intensify animal production and increase the amount of animal product turned off per hectare.

The following section explains the management adjustments that can be made to the pasture system in the higher-rainfall zone of southern Australia to increase the capacity to run more stock and to improve profit from grazing land.

Increasing pasture carrying capacity

Producers who are running farms with unproductive pastures improve their lot by addressing three issues. The first is ensuring that the soil nutrient status is adequate and that there are no major soil fertility constraints. The second is ensuring that there are productive pasture plants in the paddock that will respond to the high soil fertility and produce more feed. Finally, there needs to be sufficient stock in the paddock, and the grazing needs to be managed to ensure that the stock eat the extra feed.

Ensuring adequate soil nutrient supply

In most soils in the higher-rainfall regions of Australia the undersupply of certain key nutrients, such as N and P, is a major constraint to pasture growth. Deficiencies of other nutrients such as K, S, and molybdenum (Mo) can also limit growth. Given that N supply is provided by N_2 fixation in the nodules on legume roots, the next major nutrient whose supply is crucial to pasture production on most soils is P, with production increasing directly as the P status of the soil improves. This is because the pasture legume has a high P requirement to grow and to fix N_2 in its roots. The legume has a relatively limited root system in the surface soil and is relatively inefficient at taking up P from the soil, compared to grasses. Higher levels of available P in the soil solution, achieved by repeated P fertiliser additions, are therefore necessary for vigorous legume growth under grazing, and for the improved growth of perennial pasture grasses.

A generalised relationship has been developed for the productivity of a legume-based pasture, in terms of relative pasture DM production, and the P status of the soil. The soil P status can be assessed by taking a soil test and measuring plant-available P. A feature of this relationship is that there is a diminishing production response as the soil P status increases from low to high, or as P supply to pasture plants is progressively increased.

The message here is that if a farmer with low-P soil wishes to increase pasture production to a higher level, then it is necessary for the P status of the soil to be increased. High rates of P fertiliser (perhaps 20–50 kg P per hectare) are required to achieve this in the short term. This can be considered as a capital application of fertiliser and as a major long-term investment in the productive capacity of that pasture soil. In addition to this input, there will be regular losses of P from the grazed pasture that are related to the number of grazing animals. There will also need to be regular maintenance P applications, which should be increased as stocking rate is increased.

Managing other nutrients for intensive pasture production needs addressing. This is based on the approach of adjusting nutrient status to achieve a desired production level, and then replacing nutrients removed by activities such as grazing or fodder conservation. This is currently the practice in New Zealand, but in Australia the strategy is directed towards avoiding deficiencies. Soil testing is possible to assess the soil nutrient status for the macro-nutrients K, S, calcium (Ca) and magnesium (Mg), while tissue analysis of herbage is usually undertaken to measure the micro-nutrient status of the pasture. Nitrogen fertiliser use on pastures in southern Australia is increasing, particularly on dairy pastures. The pasture growth responses to N are quite predictable if there is adequate rainfall, and the use of N fertiliser is based on the need for extra feed, and the extent that the returns from the extra feed will cover the costs of the N fertiliser.

The need for responsive pasture species

If a major investment in improving soil fertility is to be made it is important that improved, responsive pasture plants are present in the sward. On flat to undulating arable land this generally involves the establishment of introduced legume and perennial grass species. The legume is needed because it fixes N_2, and the increased herbage N in the legume is redistributed back through the excreta of the grazing animals to the pastures, resulting in enhanced growth rates by grasses in the sward. A second benefit from the legume is its high nutritive value, such that animal performance is related to the amount of legume in the sward. The choice of legume depends on annual rainfall: subterranean clover (*Trifolium subterranean*) is used for the 450–750 mm rainfall zone while white clover (*Trifolium repens*) is the clover for the higher-rainfall regions (>800 mm). These legumes have a low, prostrate growth habit and the growing points at the end of their stolons are close to the ground surface, which increases their tolerance to heavy grazing pressure.

Introduced perennial grasses are responsive and productive in fertile soils and tolerant of heavy grazing pressure. They also have longer growing seasons than the native perennial grasses and improve the stability of production in poor seasons. The choice of grass depends on the rainfall and soil type, with phalaris (*Phalaris aquatica*) and cocksfoot (*Dactylus glomerata*) being used in drier regions (450–750 mm rainfall) whereas perennial ryegrass (*Lolium perenne*) and tall fescue (*Festuca arundinaceae*) are grown in rainfall areas above 600 mm.

If there are no improved pasture species (particularly legumes) in the sward they will have to be introduced into the paddock. Successful pasture renovation techniques have been developed for this purpose. The process can be expensive, in the order of $200–300 per hectare. It involves purchasing pasture seed, herbicides to control existing plants in the paddock, fertiliser to sow with the seed, cultivation sometimes, sowing, and sometimes grazing forgone in the short term as the new pastures establish.

The key to successful renovation is success in controlling competition from the existing vegetation. The procedure begins in the preceding spring when annual grasses are reduced by 'spray-topping' (grasses are sprayed late with knock-down herbicides to prevent seed-set). Broadleaf weeds might also need to be controlled in the preceding year by herbicide or herbicide–grazing combinations. Heavy grazing during the summer and early autumn removes all plant residues from the soil surface. A knock-down herbicide is used again after existing weeds germinate following the opening rains in autumn.

An alternative option, based on high rates of fertiliser and heavy winter grazing, is possible for around one-fifth of the full cost of renovation, if the paddock being upgraded has at least 10 per cent of desirable clovers and grasses present in the sward. Here the approach is to soil test and to apply fertiliser, lime and trace elements if necessary to correct deficiencies. Then, graze the paddock to reduce dry pasture cover in the autumn. In the winter the pasture should be allowed to grow to around a height of around 5 cm, and then graze intensively with high stock numbers down to height of 1 cm. This approach needs to be repeated through the winter. Only dry stock should be used for this purpose and should also be grazed intermittently on higher-quality pasture to maintain weight.

On land with limited development potential, such as steep hillsides with shallow, low-fertility soils, landholders have found that Australian native grasses are more productive than introduced species. They can be managed to provide valuable feed for stock. If the pasture has degraded and the density of the desirable perennial grasses such as wallaby grass is low and annual grasses are abundant, then the pasture needs to be grazed heavily from mid to late spring to limit seed set by the annual grasses. The pasture then needs to be spelled in summer to increase seeds set by the perennial grasses. Care is needed with fertiliser application. Grasslands with high conservation value need to be managed without fertiliser to avoid competition from other plants. On the other hand, native pastures containing cool-season perennial grasses such as wallaby grass (*Austrodanthonia* spp.) and weeping grass (*Microlaena stipoides*) can respond to moderate fertiliser inputs to provide valuable feed while persisting well under grazing.

Grazing management: rotational grazing or set stocking

There are two broad approaches to grazing animals on pastures. The first is to practise set stocking where the animals remain in relatively large paddocks for an extended period of 6 to 12 months, and may be moved from time to time to another paddock. In contrast, rotational grazing is a method where large mobs of animals are moved though a series of smaller paddocks according to a plan. This may be based on time intervals, or based on growth stage of the pasture, where the stock are moved on to the next paddock when the grass reaches a certain growth stage. This occurs for ryegrass pastures when the third leaf has regrown following the earlier grazing. Key differences between these approaches are that with set stocking the animal decides what it will eat and when it will eat it, whereas with rotational grazing the land manager decides what the animal must eat and for how long it has to eat it.

A number of grazing trials in the late 1990s provided increased understanding of the benefits from these grazing methods. Perhaps the main benefit from rotational grazing in high-rainfall areas is that around 10–20 per cent more stock can be run on the pasture paddocks as a result of higher pasture growth and better utilisation. This will increase animal production per hectare. Another benefit is that when there is a shortage of feed in winter, then rotational grazing allows winter feed rationing to build up feed supply, which may be invaluable if cows or ewes are to calve or lamb in late winter.

Rotational grazing will also improve the density and persistence of perennial grasses like phalaris; such persistence is a vital consideration when investing in perennial grass establishment. In addition, rotational grazing will ensure that all pasture in larger paddocks is eaten evenly, which is not likely under set stocking.

Finally, rotational grazing can deliver environmental benefits such as a substantial increase in pasture cover at the end of a dry autumn, compared with set stocking, and a more even spread of dung and urine across the paddock. This is particularly important with Merino sheep as they like to camp together in the evenings under set stocking, resulting in the accumulation of dung and urine in sheep camps.

Rotationally grazing small paddocks also increases the farmer's control over the amount of pasture available to animals, and the intensity with which a particular pasture is grazed, compared with set stocking. Take, for example, the dairy farmer who is able to utilise pasture effectively on the farm. The farmer is able to regularly walk across each paddock and measure the amount of available pasture, and calculate pasture growth rates (the differences in pasture mass between walks, divided by the number of days between walks). The calculation will enable the farmer to decide when paddocks will be grazed by the herd, and if there is going to be enough feed for the cows in a particular paddock tomorrow, or later in the month.

Set stocking has some advantages. Animal performance per head will generally be better under set stocking for a number of reasons. (Note though that animal performance per hectare is the source of profit.) The animal can graze younger pasture, as the plants have not been spelled. There is likely to be more clover in the set stocked paddocks, and clover is more nutritious than grasses. In the more relaxed set stocking regime the animals have more opportunity to select the highest-quality parts of the pasture plants. In addition to the better per-head performance of animals in set stocked paddocks, there is also less cost in labour as the animals do not need to be moved from paddock to paddock on a regular basis.

Changing the time of lambing or calving

Ewes give birth to their lambs in the autumn on many farms in south-eastern Australia. This means that the ewes encounter dry, low-quality pasture during early lactation, when their need for energy is at a maximum. Invariably this means that either (1) the farm's stocking rate is limited to the number of animals that can be carried successfully through the autumn and winter, or (2) that supplementary feeding will be required to ensure that there is no feed shortage. Changing the time of lambing to the early spring means that the peak demand for feed coincides with the time of peak pasture supply. Furthermore, the ewes have a lower feed demand in the winter, when pasture growth rates are at a minimum due to cold winter temperatures. This means that more

FIGURE 12.5 Eleven-month-old (plus) Merino weaners that were born in late winter, after a mild summer. The farmer made the switch from autumn lambing to late winter several years earlier. The farm was immediately able to increase stock numbers by 30 per cent as lambing ewes did not need to be fed during the winter.

ewes can be carried successfully through the winter with spring lambing as opposed to autumn lambing, which means a higher overall stocking rate for the farm. The matching of feed supply with demand in this way leads to fewer pasture surpluses in the spring or deficits in the winter. Similar outcomes occur when the time of calving is changed from the autumn to the late winter – early spring, although the impact is less with cattle because of their longer gestation period.

There are real possibilities for increasing the carrying capacity of the farm by changing the time of lambing and calving, to avoid having to manage lactating females in the autumn. There will be the consequence of having to deal with young stock, which are weaned from their mothers, and then encounter poor quality summer or autumn feed. Young Merino weaner sheep are an example: they have no body reserves, and high weaner mortality will occur unless they have access to high-quality feed to maintain growth rates. Indeed, weaner ill-thrift has been a major reason why farmers in southern Australia choose not to lamb in the spring. Thus special strategies, such as access to special purpose pastures or supplementary feeding, are needed with spring lambing and spring calving to ensure that sheep and cattle weaners are fed adequately in the autumn months.

Achieving animal performance targets

The extent to which grazing animals perform well on pasture, in terms of their growth rate, or wool or milk production, depends on whether they can obtain sufficient energy from the pasture to meet their needs. There are two ways to increase energy supply for the animal from pasture. The first is to provide high-quality pasture in the form of green, leafy pasture that has legume growing in it. Animals are able to eat more high-quality pasture and each mouthful of this pasture has higher energy value, compared to a low-quality pasture. The second is to increase the amount of this pasture in the paddock; this enables the animal to take full mouthfuls of pasture and to select the highest quality from the pasture mass. Farmers know that animals with high energy demands, such as cross-bred ewes with twin lambs at foot, must have access to tall (>5 cm high), dense, high-quality pasture if the ewes are to provide sufficient milk for the twin lambs to grow rapidly.

As pastures mature and senesce in the late spring – early summer, their quality declines appreciably. Green leaf gives way to dry stalky material. The animals cannot eat much of this material as it fills up their first stomach, and it remains there taking time to digest. Only a small part of the material is digested and so the animal cannot acquire much energy from this poor quality feed. This means that animals with high energy demands such as lactating females or young growing stock cannot perform on these pastures. They are only really maintenance pastures, and, indeed, it is likely that dry stock will lose weight on such pastures. The amount of this dry feed in the paddock declines over the summer, further reducing the opportunity for animals to select higher quality material from the pasture mass.

There are certain groups of animals that are vulnerable to energy deficits and significant production penalties occur if they are forced to graze low levels of poor quality pasture. For example, the breeding ewe in late pregnancy needs 75 per cent more herbage (150 per cent more in twin-bearing ewes) than the non-pregnant animal. Severe underfeeding during late pregnancy may result in ewe mortality from pregnancy toxaemia. High lamb mortality will also result from poor mothering by the ewe, and will be exacerbated by low lamb birth weights. Continuing undernutrition in early lactation will further threaten lamb survival due to low milk production; in addition, wool growth by the ewe, the tensile strength of her wool and her future reproductive performance may be reduced.

Climatic variability—the need for management resilience

One of the features of the Australian environment is the relatively unreliable rainfall. All indications are that climate change will mean that future rainfall is likely to be even more unreliable. This presents real problems for the farmer who is attempting to increase the carrying capacity of his/her pasture land and to run high sustainable stocking rates. There is a need to ensure that stock numbers are high enough so that good profits can be generated in the good (high-rainfall) years. On the other hand, in the dry years, when the autumn break is very late, and worse still if the spring rainfall fails, and a drought occurs, then the farmer needs to avoid the worst case scenario. This is when there are large stock numbers on the farm, no feed in the paddocks, collapsed prices for the stock, and very high costs for feed supplements. Risk management plans enable the farmer to manage climate variability effectively to maximise profits in the good years and minimise losses in the bad years.

Management resilience in response to climate variability requires a strategy plan with critical trigger points. When these points are reached, firm pre-determined decisions need to be implemented. Consider, for example, a grazing property in southern Victoria with a 700 mm rainfall and an average 7-month growing season, running a conservative sustainable stocking rate of 15 DSE/ha. An early autumn break in mid March with follow-up rain in April provides a trigger indicating that there will be above-average growth of pasture in the autumn enabling the farmer to run more stock perhaps in the form of agisted dairy heifers or trade steers to grow out over the growing season. An autumn break in mid May means that feed supply over the winter will be less and so some mobs of sheep such as the cast-for-age ewes, and cull weaners, need to be sold forthwith. The difficult trigger occurs when rains come in late June, which means that feed supply will be critical. Pasture growth will be delayed and growth rates will be very low due to cold temperatures. Extra supplementary feed must be purchased and more stock will need to be sold.

The situation can become still more critical if there is no rain in August. Increasing temperatures and sunny days in August will result in pasture growth that depletes soil water. Should the Southern Oscillation Index (SOI) during the winter months, which is a useful indicator of spring rainfall, be negative, there is a further indicator that spring rainfall will be low. The farmer then has to take further drastic action: more lower-priority stock need to be sold before their price collapses, and more feed supplements need to be purchased before the prices escalate, as the prospect of a full-blown drought resulting from a failed spring rainfall looms.

These are difficult times for the farmer, as it is easy to think that it might rain in the next ten days, which will make all the difference. Real discipline is required to stick to the plan and to take decisive action when the trigger points are reached. Delaying action will lead to lower stock prices and high feed costs, should the drought set in. The situation is easier if some of the accumulated profits from the good years are set aside in drought management bonds, and these can be used to purchase the feed supplements that are increasing in price.

The specific details given in the text above will not apply to all farming businesses. Different options for expanding stocking rate and animal output in good years, and for decreasing these settings in bad years, will vary for different property types and enterprise mixes. Nevertheless the need to have flexible resilient plans that respond to seasonal conditions, as early as possible, will be required for each farm. What is becoming increasingly apparent is the need to improve the accuracy of our rainfall forecasts. High priority research is underway, and must continue to be undertaken, to achieve better forecasting accuracy to enable these difficult decisions to be taken with greater confidence.

Disappointment with tropical pasture

The major form of land use in the tropics and subtropics is the extensive grazing of beef cattle on native pastures. These regions have a predominantly summer rainfall, where a relatively reliable summer wet season is followed by a dry season during the cooler months from April to November. The native pastures consist of summer-growing perennial grasses that grow rapidly following the opening rains and produce quite nutritious feed for several months. The plants are generally adapted to a short growing season, and after flowering their quality declines markedly.

During the dry season the dry feed is deficient in protein and energy, and beef cattle lose weight. These fluctuations in pasture quality reduce cattle turn-off for many years. The reproductive performance is poor with low branding percentages. Also the animals are older and heavier at slaughter than those in the south. Initially the cattle were all British breeds, such as Shorthorns, Herefords and Devons, which were poorly adapted to the high temperatures, the poor dry-season forage, and to cattle ticks in the wetter areas.

When the pasture improvement revolution was occurring in southern Australia in the 1950s and 1960s, questions were asked about the possibility of similar improvements for northern Australia. Could the use of fertiliser and the introduction of adapted tropical legumes and grasses increase animal production in the north where the tropical soils are deficient in nitrogen and phosphorus? A new Commonwealth Scientific and Industrial Research Organisation (CSIRO) Division of Tropical Crops and Pastures was established and promising tropical grasses and legumes were introduced from overseas. Major research in tropical and subtropical pastures continued during the next thirty years.

The results from the large effort to improve pastures in the tropics and subtropics have been disappointing, though the introduction of *Bos indicus* × *Bos taurus* cross-bred cattle has had a profound impact on productivity. In the drier regions (500–750 mm rainfall), the clearing and improving of brigalow lands were successful. Here the low forest communities dominated by brigalow (*Acacia harpophylla*), which grew on relatively high-fertility clay soils, were mechanically cleared and sown down to predominantly buffel grass (*Cenchrus ciliaris*) pasture. Apart from the brigalow the move to widespread pasture improvement has not occurred. A key aim was to get vigorous persistent legumes established to improve the pasture's protein concentration during the dry season. Many of the legumes did not persist. Siratro (*Macroptiilium atropurpureum*), which was released by the CSIRO in 1962, failed to persist under heavy grazing, which could not be avoided at times. Other twining and scrambling perennial legumes including *Centrosema* sp. and *Desmodium* sp. were also sensitive in varying degrees to continuous grazing. The promising annual legume, Townsville stylo (*Stylosanthes humilis*), was sown and spread over large areas, but it succumbed to the fungal disease anthracnose in the 1970s, and no longer became a viable option.

The lack of a suitable legume is one problem of the buffel grass pastures that have been established on brigalow land and on the less fertile forest country in central Queensland. Nitrogen deficiency becomes the main limitation to production as the nitrogen reserves in the soil decline over time, in the absence of an adapted legume.

Despite these disappointments, there has been some progress. Alternative disease-resistant *Stylosanthes* species have become available for the oversowing of native grass pastures. In addition, a number of special purpose pastures such as ponded grasses and stands of the tree-legume *Leucaena* have been developed in certain environments, which have increased animal output significantly.

Past mistakes in the arid pastoral zone

Pastoralists with their sheep flocks settled the dry, arid inland regions during the final decades of the nineteenth century. Originally the lack of water prevented settlement. The discovery and use of artesian wells from 1879 onwards, together with the establishment of earthen tanks and networks of bore drains, made a permanent pastoral industry possible in the arid inland. As it happened, the timing of this pastoral occupation proved to be a disaster, due to a run of years of above-average rainfall, particularly in the 1880s when rainfall was exceptional. This, allied with the presence of plentiful native grass species in many areas, created a misleading impression of what a viable stocking rate might be over time. Stock numbers in the arid zone increased substantially.

The first landholders did not (and could not) know that the recent seasons had been exceptional and that the land was stocked above capacity. They had little understanding of the variability of the Australian climate. A period of widespread drought began in 1895 and reached extreme severity in 1902. The consequence of increased stock numbers, and the competition from rabbits, which had reached plague proportions in the inland, led to the degradation of the countryside. Starving and thirsty animals died in their thousands.

Significant land degradation and vegetation decline occurred during this brief period of land occupation. Severe wind erosion took place as huge dust storms occurred during the latter stages of the drought. The source of much of the airborne dust was claimed to be the arid zones. Subsequent water erosion occurred on the bare landscape. In addition, many useful native forage plants were permanently lost, and this reduced grazing capacity of the arid lands. The reduction in plant cover led to a cycle of soil displacement by raindrops, surface sealing and increased run-off and topsoil loss, increased soil temperatures, the loss of soil organic matter and a reduction in soil aggregate stability. The general effect was the creation of harsh conditions for seedlings, limiting the revegetation capacity in wet years.

More than a century has now passed since the arid grazing lands were damaged by grazing animals. Grazing still continues as the major form of land use. There is now much interest among pastoralists as to how they might improve the condition of the arid lands, and the grazing achieved from them. There are trials underway to test different grazing strategies that spell pasture between periods of grazing. The aim is to develop grazing management, paddock size and water-distribution guidelines for more sustainable land use in the arid zone.

Improving the profitability of mixed farms

Mixed farms are those with a mix of cropping and livestock enterprises on the one farm. They are very common across the Australian wheatbelt and in the new high-rainfall cropping regions. There are many reasons for combining these enterprises on the farm. They might relate to the capabilities of the different land types on the farm, or to the need for a diversified income. The extra management complexity is justified by the perceived benefits for the overall farm business. Ideally there is a synergism between the enterprises, where one will benefit the other, and vice versa. This section describes some of the ways that the different enterprises impact on each other, and outlines new developments and opportunities for improving the performance of the mixed farm.

Grazing pasture phases in the cropping rotation

The use of short-term pasture phases of three to five years became a widespread practice in cropping rotations in the southern cereal belt after the Second World War. These legume-based pasture phases played a very valuable role in maintaining soil fertility in the southern cereal belt during the 1950s, 1960s and 1970s but they have declined in productivity, with lower legume content and more annual grass weeds, which act as hosts for cereal root diseases.

Further, the relatively high grain prices and improving yields meant that cropping became more profitable than livestock enterprises. This resulted in many farmers cropping continuously, using grain legumes and fertiliser nitrogen to replace the nitrogen supplied by the pasture legumes, and to switch to minimum tillage practices to attempt to maintain the physical condition of the soil. Nevertheless, it is probable that soil physical properties and organic matter levels cannot be maintained by these practices, and a return to pasture leys may be necessary to maintain soil fertility over the longer term—if only the returns from the grazing enterprise improve.

Increasing use of grazing lucerne in the rotation

One approach to improve returns on mixed farms is the replacement of annual pastures in the rotation with the perennial legume lucerne (*Medicago sativa*). For example, an economic analysis of 13 farms in northern Victoria compared the effect of replacing annual pastures with lucerne. The analysis found that the average **net present value** of the farm increased by 35 per cent with lucerne. The basis for the increase was the profitable increase in the livestock enterprise, associated with large increases in stocking rates, with a focus on sheep meat production rather than wool production. There were minor improvements in crop profitability.

FIGURE 12.6 Cross-bred ewes and prime lambs grazing lucerne in the spring

Lucerne can provide a number of unique advantages to the cropping system. This is not surprising as the legume is a special plant: it is the most widely grown forage legume around the world and has been used by human kind for thousands of years. The plant has a deep tap root that enables it to use more water from deeper soil layers than annual plants. The plant can dry out deep subsoil layers and this has significant advantages in managing dryland salinity, reducing waterlogging and improving root growth in dense clay subsoils. Other benefits come from its N_2 fixation, which results in high grain protein in following wheat crops, and lower N fertiliser requirements for these crops. Its perenniality enables it to respond to out-of-season summer rain, and allows for easier control of winter and summer weeds. There is also the option for cutting the lucerne at flowering and making quality hay, which can provide additional income.

As with most changes to a cropping system, using lucerne presents challenges for farmers. In the dry Mallee region establishment failures occur because of dry conditions. Another difficulty is the reduction in the yield of the crops that follow lucerne in dry years, due to the dry subsoils. Other farmers have developed effective strategies by removing lucerne earlier in the preceding spring and putting a chemical fallow in place to store spring rainfall in the soil for the following crop. Another problem is the slow winter growth by lucerne, which contributes to a feed shortage in the winter. This problem has been overcome by sowing a grazing cereal crop into the near-dormant lucerne, following a herbicide spray to suppress any lucerne growth. The cereal can be grazed through the winter and into the spring; it will be eliminated when the reproductive apices in the elongating stems are eaten off by the stock.

Another innovation for mixed farms is to grow a pure annual legume stand for one year, followed by a winter crop in the second year, then allowing the legume to regenerate in the third year, with another crop in the fourth year. This has been developed for farms in the medium- to high-rainfall region of southern Australia where self-regenerating legumes such as subterranean clover are widely used. Once the system is established the only inputs are the crop seed, some herbicide to suppress the legume seedlings in the autumn, and nutrients such as P, K and S as required for maintaining crop and pasture production. There is no need for any N fertiliser. There will be sufficient hard seed in the legume seed bank in the soil for the legume to establish in large numbers in the second year after the cropping year. The viability of this approach depends on profitable returns from the clover phase; the crops yield exceptionally well after the legume phase.

Grazing crops within the cropping rotation

Further variations within the cropping system are possible where animals are allowed to graze on cropping paddocks, within a continuous cropping program. This can either occur after the crop has been harvested, or early in the vegetative stages of crop growth if special dual-purpose crop varieties are grown for the purpose.

Grazing crop stubbles

Moving stock onto the stubble paddocks after the grain has been harvested is a very widespread practice. The animals are able to clean up unharvested grain and any weeds or crop regrowth that might be present in the paddock. Sheep are more suited to this task than cattle as they can graze closer to the ground. Stubbles therefore provide very useful 'grazing days' for animals on mixed farms. The feed quality of the stubbles is low so the animals will begin to lose weight once the grain and any green growth have been consumed. The animals perform the task of controlling weeds that might grow in the stubble paddocks, and so savings on herbicide costs are possible.

Despite the apparent advantages for both enterprises, many farmers are choosing not to run stock on their stubble paddocks. These are the crop farmers who find labour supply to be limiting, or have adopted minimum or zero-tillage and stubble retention practices, together with controlled traffic where the machinery is confined to permanent tracks, to minimise soil compaction. The farmers consider that stock result in treading damage and soil compaction—damage that they are trying so hard to avoid. Further, the animal movements across the paddock tend to bury any weed seed into the surface soil, prolonging weed seed viability. The decision to remove stock from cropping paddocks, and from the farm, can be long-lasting if fence lines are removed and watering points become neglected. Such decisions are made on the basis that improved water use efficiency and more favourable soil structure will deliver greater benefits than those from the diversified enterprises on the mixed farm.

Dual-purpose winter crops

New opportunities for mixed farmers in the higher-rainfall regions are provided by dual-purpose wheat crops. The approach here is to sow long-season wheat varieties early in March or April. The early wheat can then build up a mass of vegetative growth that provides very useful, high-quality winter grazing. So long as the animals are removed from the wheat paddock before the stem elongation stage at the end of tillering, then there is not likely to be any major penalty for grain production. The dual-purpose wheat strategy seems to offer a real bonus to the mixed farmer in the form of (1) valuable winter grazing at the time of the year when feed on the farm is scarce, plus (2) the grain from the crop at harvest. However, there are uncertainties with this strategy. The first is the requirement for early sowing rains to establish the crop in the early autumn. The second is the possible threat from wheat streak mosaic virus, which is spread by the minute wheat curl mite. Severe damage occurred in early-sown dual-purpose wheat crops in all eastern states in 2005. It remains to be seen whether this disease develops into a major constraint for these crops in the years ahead.

In addition to the dual-purpose wheats, there are also dual-purpose canola varieties that show promise. Early trial work has shown that winter canola varieties can provide valuable winter grazing and recover and go on to produce high grain yields with satisfactory oil concentrations in the grain. The canola needs to be sown in April, with grazing occurring during the winter months until the end of August. More research is required to determine which varieties are most suitable for this dual-purpose role, and whether the canola can recover from grazing and produce high grain yields, at least equivalent to ungrazed canola, in different years with wet or dry springs.

Improving the genetic potential of animals

The term *improving animals* refers to the achievement of better levels of performance in some defined animal trait, such as growth rate or reproduction rate or clean fleece weight or average fibre diameter, or carcass weight and composition. There is one very important point to make at the outset: improved genetic potential of animals is not a 'free lunch'. Translating improved potential for performance into actual improved performance requires other changes in the animals' environment. This involves further investment of capital and labour. Animals with the potential to produce more output need extra feed, or removal of environmental factors that are limiting performance in order to fully express their genetic potential.

Selection criteria are traits of an animal that the breeder wishes to perpetuate in future generations. This involves selecting objectives that are responsive to selection and will contribute to greater profit. The more traits selected, the slower the improvement in the achievement of any one trait. In order to select according to a trait breeders use estimates of an animal's breeding value from measurements of that trait in the animal itself and from measurements made on its relatives. Many of the characteristics of interest such as growth rate and carcass quality are of medium to high heritability. The performance of an animal is a good guide to the genes carried for these characters. For characters with low heritability an animal's performance tells little about the genes for these characters carried and passed onto offspring. As an individual animal has some of its genes in common with its close relatives, more information can be gained about genetic value of an animal for low heritability characters by examining the performance of its relatives. This is the function of progeny and sib testing.

Traits in animals are often correlated. Measurements on correlated traits can be used to improve the estimate of an animal's breeding value for a related trait. Using statistical techniques, best estimates are made of an animal's breeding value for each breeding objective. Breeding values achievable in given environments need to be converted into economic values in order to derive a selection index. The estimated breeding value (EBV) for an animal's genetic merit can be estimated by detailed analysis of performance records. EBVs are expressed in physical units and as differences from a population average (flock, herd, breed). Reproductive characters are difficult to improve by selective breeding because the heritabilities of reproductive characteristics are low. On the other hand, the higher heritability characteristics are fairly easy to improve by genetic means, particularly the growth characteristics. This is done by selecting herd replacements on pre- and post-weaning growth measurements of young animals.

Inbreeding occurs when breeding is from individuals that are more closely related than the average closeness of relationships in the population. Inbreeding increases homozygosity where traits, good and bad, are magnified and cause a loss of vigour. The degree of inbreeding causes a similar decline in reproductive performance. Linebreeding is a form of inbreeding where a particular sire or dam's genetic merits are perpetuated.

Mating individuals that are not related, such as crosses between breeds or strains, generates heterosis or hybrid vigour in the offspring. The benefits of cross-breeding are well established and widely exploited in the fat lamb, pig and poultry industries. Cross-breeding is not so widespread in the cattle industries. The hybrid vigour (heterosis) effect in the cross-breed, in the case of cattle, shows itself in higher calving rates, calf survival, weaning percentages, weaning weights, and post-weaning weight gains.

Cross-breeding can be simple or rotational. Simple cross-breeding involves using a terminal sire and no progeny are retained. Replacement purebred females have to be bred separately or bought-in. Rotational cross-breeding involves using several breeds and crossing them in a cyclical way over time: for example cross two breeds to produce a first-cross animal, then mate the first-cross females to the breed of the sire used (two-breed cross-cross), or to a third breed (three-breed cross-cross). Management of cross-breeding programs can become complicated. Still, while the advantage of the cross for each production characteristic may be only 3 to 4 per cent, their effect is cumulative: that is, more calves are born, these calves survive better, more of them are weaned and at a higher weight, and after weaning they grow better. The cumulative effect of these advantages can be to give a production increase that is bigger and quicker than can be obtained by selecting within a breed. Cross-breeding of British with Brahman (*Bos indicus*) types not only gives hybrid vigour but also allows other production characteristics to be introduced into the progeny, for example better heat tolerance and tick resistance.

As the genetic worth of the animals changes through breeding, different environmental limiting factors come into play. Environmental factors continually constrain performance. For example, it is often thought that by breeding animals that have superior genetic potential to produce more output, the extra output is somehow 'free'. That is, the extra output is produced by animals that would eat the same amount of feed, and require the same husbandry and costs as inferior animals that produce less output. This is not the case. To achieve their greater genetic potential for production, the generally genetically superior animals need to eat more food than the lower producing animals. The exception is for animals that are genetically superior in net feed efficiency and conversion of feed to muscle. There are increased costs associated with increased output. The production of genetically superior animals does not increase unhindered simply because of improvement in one part of the animal's potential capacity. Instead, as production increases occur, new environmental constraints are met. The true benefit from genetic gains in particular production traits that appear theoretically possible is the value of the total benefits minus extra costs.

Improving livestock husbandry

Improved livestock husbandry practices have been adopted by most domestic animal enterprises in recent years. An example has been the development of improved practices for the beef herd in tropical northern Australia. More than 50 per cent of the total Australian cattle population are located in the north of the continent. Traditionally, about 80 per cent of the beef produced in the north was destined for export as heavy-weight ox, where it took from four to six years for the animals to reach market weights. Changing market conditions, such as access to Japanese and some other Asian markets and their demand for high-quality beef, and the growth of the live cattle trade, have resulted in a trend towards the production of younger-age cattle to meet specific markets.

The major task is to improve growth rates to reduce the age and beef quality of animal turn-off. Major improvements in production have occurred with the widespread use of better-adapted *Bos indicus* genotypes and cross-breeding through the north, and these types now account for over 85 per cent of the northern cattle population. Businesses with inland properties often grow out young stock on better-grassed coastal properties and finish them in feedlots. Also, improving pastures in the north by oversowing native pastures with different tropical legumes such as *Stylosanthes* species, discussed in the previous section, is a major benefit.

Reproduction efficiency has been a major problem in northern herds. There are now new management strategies available to improve the reproductive performance in northern cattle. The historical average weaning percentage for northern herds has frequently been less than 50 per cent. Increasing this is a major source of improvement in beef productivity in the region. Strategies to bring about this improvement include (1) effective heifer segregation to ensure that the first mating and conception occurs as two year olds with body weights between 280 and 320 kg; (2) preferential treatment of first-calf cows to improve survival and fertility; (3) managing to ensure that the cow's body condition is maintained through strategic weaning and dry season supplementation with either molasses-urea or cottonseed meal; (4) appropriate nutritional care of early-weaned calves to minimise mortality problems; (5) objective culling of cows based on infertility, temperament and age, and of unsound older bulls or younger bulls that are entering the herd.

Another area where improvements can be made is in labour costs: the single most significant cost for beef producers in the north. A number of automated cattle management technologies to minimise labour requirements for mustering and cattle handling have been developed. Examples such as the use of spear gates to trap cattle on water or with lures such as hay with phosphorus supplements, are now widely used in the north. Automatic drafting devices such as the use of a cow/calf separator in the form of a modified spear gate assist the weaning process, while automatic medicament applicators are available to apply chemicals for tick, buffalo fly or worm control. These devices are tools that assist in the adoption of important strategies such as early weaning and developing improved nutrition and disease control management outcomes.

More recently, improvements in livestock husbandry have focused on improving product quality and individual animal efficiency. To achieve these goals several strategies have aimed to improve both quality assurance and animal welfare at all stages of the production and processing chain. Depending on the product in question, this often includes understanding consumer perceptions as part of the strategy. A good example of quality assurance has been the development of the Meat Standards Australia grading system. This is a beef and sheep meat eating-quality program that labels beef and sheep meat with a guaranteed grade and recommended cooking method. It incorporates recommended guidelines for a wide range of cattle and sheep management practices, processing systems, cuts, ageing periods and cooking methods to optimise the impact each has on eating quality. Similarly, animal welfare has improved by the development of training packages and guidelines to improve stock handling practices at all stages of production from on-farm through to abattoir processing. The aim here is twofold. Animal well-being is better, and by reducing the negative impact of a range of human and environmental stressors on animal production, performance and product quality is also enhanced.

Efficiency of animal production is also improved by increasing nutritional knowledge. For instance, recent research has identified specific nutrients and feed stuffs that can (1) reduce greenhouse gas emissions from ruminants, (2) alter the functional food properties of milk, red meat and fat, and (3) maintain good animal health and therefore production (for example by reducing the impact of internal parasites and heat stress).

Improving intensive plant production

Irrigation

Essentially, irrigation involves improving systems of delivering water to plants. Good irrigation involves getting the slope of the land right. This means water moves down through the soil profile and laterally across the soil area at a rate appropriate to the aims of meeting requirements of plants and ensuring excess water is drained quickly from the plant root zone and does not add to the deeper groundwaters.

Many developments in spray and trickle systems of water delivery facilitate more efficient water use by enabling controlled and precise applications of water to individual plants, such as devices to measure soil moisture and micro-sprays. These innovations dramatically reduce the water used to grow plants, and reduce water wastage and the problems that follow from excess water put on land.

The main technical advances in irrigation have been reducing the labour component involved in applying water to pasture and crop. Labour is replaced by capital investment in plant, structures and timing devices designed to make the process more automatic. Devices indicate when pastures and crops require irrigation so that yield is maximised. Nitrogen is used on pastures and crops to exploit favourable temperature and moisture conditions. The superiority of rotational grazing of irrigated pastures over set-stocking at high (by district standards) stocking rates is increasingly recognised.

Investment in irrigation projects with good reliability to supplement rainfall can be profitable when good-quality feed production is low on the rest of the property, provided that the feed is used in activities that have a high marginal return per hectare of irrigated pasture: for example finishing lambs or calves for an out-of-season market, and growing heifers or hoggets to calve or lamb at two years of age or younger. The value of a unit of feed varies according to the season of the year, because it has a range of effects on the total gross margin and profit. In the same way, the feed produced by irrigation has different effects on total gross margin, and profit, as its use will give a wide range of marginal returns.

Integrated pest management

In recent years there has been a major move towards reducing the reliance on chemical pesticides for managing pests on Australian farms. The repeated use of pesticides increases the risk of their eventual demise because the target pest develops pesticide resistance. There is also an increasing worldwide trend in consumer preference for food products that are produced with minimal use of chemical pesticides. The new strategy is Integrated Pest Management (IPM), where a variety of management tools, including strategic pesticide use, are employed to control the pest.

IPM involves applying the whole farm approach to decisions about pest control in crop, fruit and vegetable activities. The IPM approach is based on a range of ways of tackling pest problems. It differs from traditional approaches that focused narrowly on chemical control of pests. The IPM approaches are characterised first by combining a variety of pest control methods (cultural, chemical and biological). Second, the status of the crop and pest population is evaluated empirically by regularly monitoring the growing conditions of the plants. Third, actions are taken when pest numbers reach critical levels and the benefits of taking particular actions are likely to exceed the costs. The strategic approach to pest control that characterises IPM methods reduces costs of pest control, usually from less chemicals being applied to crops, fruit and vegetables.

Productivity growth and agricultural research

Growth in the economy, and reductions in world hunger, come from increasing productivity faster than the population grows. Investment in research and adoption of innovations by farmers are major sources of improvements in productivity.

Productivity growth

A key to the continued profitability in Australian farming has been the rapid growth in productivity on farms. For agriculture as a whole, productivity growth over the period 1965–66 to 2008 was 2.5–3 per cent per year. Productivity gain was around 1 per cent per year for the economy as a whole. Over the past thirty years the output per unit of capital in agriculture has increased substantially and the number of persons employed full-time has fallen, that is, the productivity of capital and labour has been rising.

Productivity gains come from rural research providing improved techniques, as well as from larger farms (economies of scale), better farm management, better educated and informed farmers, and cheaper and better inputs. Farmers also benefit from potentially large efficiency gains beyond the farm gate, such as in processing, and in transport and handling. Australian agriculture's growth relative to that in manufacturing and other industries has generally been much higher than in most developed economies. Thus farmers have been in a good position to compete with other industries for resources. In recent years productivity in overseas agriculture has been increasing as well. This adds to competitive pressures on farmers.

Agricultural research

Declining investment in agricultural research in the past decade has led to slowing of growth in world food supply. In a world of rising population, improving productivity of agriculture is the major objective of agricultural research. The benefits from agricultural research come from reducing costs of producing a given quantity of output, or, enabling more output to be produced for a given total cost.

The benefits of research are assessed using **benefit cost analysis** of a research project. In a benefit cost analysis of agricultural research an attempt is made to measure the benefits in terms of the value of increases in production minus the cost of increasing production and the cost of the research work itself. Most of the estimates of benefits from rural research show high rates of return. This does not mean that every research project in agriculture has a good return, as research is an activity with highly uncertain outcomes. One successful research project may have high returns, but it may be that many do not. Knowing something is not a profitable avenue to pursue is also useful if it prevents someone else trying it some time later.

Applying research findings to farming

Agricultural science is an applied science concerned with providing data necessary for an informed decision as to what should be done. Farmers have a range of goals, but always economic objectives are included and often predominant. Therefore, analyses to provide recommendations for farmers need to be based on criteria of economic efficiency (for example net income) rather than technical efficiency (for example maximum output per animal or per hectare). Decisions on 'now what do we do?' will be made by applying economic and other criteria to the physical data provided by the agricultural scientist.

Inferring from an experimental situation to the situation on a particular farm requires a good deal of judgment. Research results are adjusted to farm performance, to allow for differences between experimental conditions and the conditions on a particular farm, including the quality of the farmer's management. Inevitably, the question arises: 'how useful is it to make recommendations to farmers on the basis of sample data from agricultural experiments, and what are some of the problems associated with doing so?' Experimental yields are adjusted to yields achievable on farms because:

- the assumed continuous relationship may not exist
- there is interaction between the factors
- an estimate of a **production function** applies to particular soil in a particular region in a particular year or run of years
- the managerial and labour intensity on experimental plots is different from that on the larger areas of farms
- for a given level of technology, farm yields as a proportion of comparable experimental yields tend to diminish as the size of the farm increases
- future yields and prices are uncertain.

Extension

A common lament in agricultural scientific circles is that there are plenty of good technical discoveries that farmers should adopt but have not done so. The implication is that farmers are somehow remiss and 'if only they knew', these new technologies would be quickly implemented and would 'save the day'. More often than not, the real situation is that, once all things are considered, including the farmers' goals and risk, the change in question is 'not quite right' for the situations of many of the farmers in question. Indeed there are always some farmers who already know how to farm better than they do!

Adoption of any innovation is a complex process as the rate of adoption of a change in any particular farm business depends on the nature of the change, the technical and economic situation of the farm business, and the aspirations and characteristics of the farmer involved. Expected advantages, in terms of net returns and risks, and physical operational benefits, are the keys to the decision about whether to do something differently or to continue as before.

It is simplistic to think that there are numerous innovations that promise huge rewards simply waiting to be adopted by farmers. History shows that farmers rapidly adopt changes that provide clear net gains, in terms of returns and risks, and operationally. Generally, the expected gains from change can be evaluated using the whole farm approach and calculating a few simple farm-management budgets.

REVIEW QUESTIONS

1. What have been the major steps forward and backward in pasture improvement over the past 150 years?
2. Discuss the main components of productive pasture technology that is being developed in high-rainfall temperate regions in southern Australia.
3. Extra production from genetic gain in animals does not come free—it has to be 'paid for' with extra feed. What does this mean?
4. How would you compare high- and low-input farm systems?

FURTHER READING

Davidson, B. R. and Davidson, H. F., *Legumes: The Australian Experience*, John Wiley and Sons, Brisbane, 1993.

Loomis, R. S. and Connor, D. J., *Crop Ecology: Productivity and Management in Agricultural Systems*, Cambridge University Press, London, 1992.

Wheeler, T. L., Pearson, G. T. and Robards, G. E. (eds), *Temperate Pastures: Their Production, Use and Management*, Australian Wool Corporation/CSIRO, Australia, 1987.

WEBSITES

Australasian Farm Business Management Network: www.csu.edu.au/faculty/science/saws/afbmnetwork/

Australian Bureau of Agricultural and Resource Economics: www.abareconomics.com/

CSIRO: www.csiro.au/

NSW Department of Primary Industry: www.dpi.nsw.gov.au/

Victorian Department of Primary Industry: www.dpi.vic.gov.au/dpi/index.htm

Changing Climate

13

CONTENTS

Introduction

There is now sufficient evidence that the global climates are changing, with the mean surface temperature of the earth warming by 0.9°C in the last one hundred years (IPCC 2007). Global sea surface temperatures have warmed by about 0.5°C. Eleven of the 12 years 1995–2006 rank among the warmest years in the instrumental record of global surface temperature (since 1850). Scientists believe that these changes to the world's climate are due to the increased concentrations of greenhouse gases carbon dioxide (CO_2), methane (CH_4) and nitrous oxide (N_2O) in the earth's atmosphere. These major greenhouse gases are responsible for 98 per cent of the climate change over the past one hundred years. The increases in the greenhouse gases are due to human activity: CO_2 from the burning of fossil fuels (75 per cent) and deforestation (25 per cent), CH_4 from ruminant farm animals and the cultivation of rice, and N_2O from the use of nitrogen fertilisers.

FIGURE 13.1 Global and continental temperature changes in the last century across all continents and oceans. Black lines represent temperature difference (anomalies) from the 1960–90 reference periods (IPCC). The dark shading represents the modelled temperatures with all natural phenomena included (volcanoes, sun's activity and changes in the earth's axis); the light shading represents all natural phenomena with greenhouse gases included.

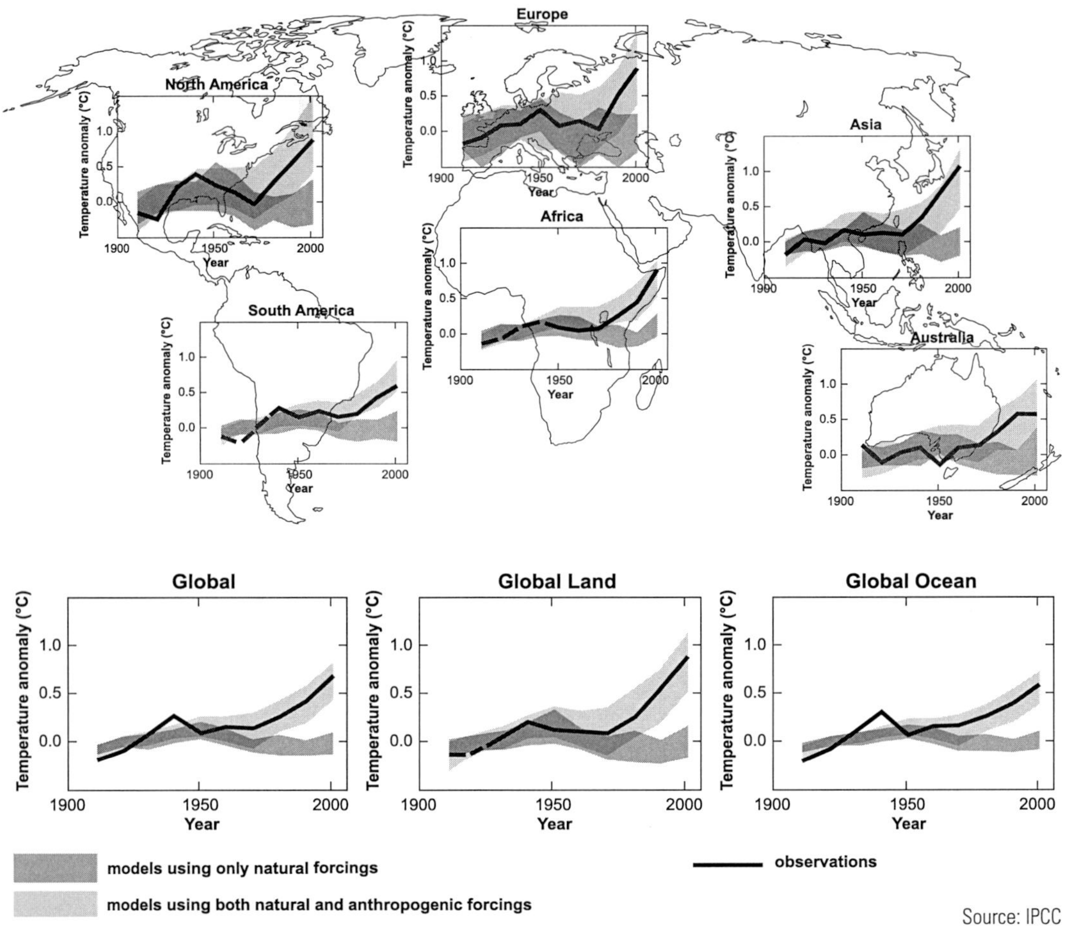

Source: IPCC

The recent Intergovernmental Panel on Climate Change (IPCC) fourth assessment report clearly indicates that all continents have warmed significantly in the last half century and particularly in the past 25 years (IPCC 2007, figure 13.1). The northern hemisphere continents, North America, Europe and Asia, have warmed more than the southern hemisphere continents. The land has warmed more than the oceans. Significantly, this global warming cannot be adequately explained by natural phenomena such as volcanic activity, sunspot activity and changes in the earth's axis (IPCC 2007). When greenhouse gases are included in the models the observed global warming can be explained (IPCC 2007).

As climate is integral to any agricultural enterprise, this chapter examines the way climate has changed in recent times. The implications of climate change for agriculture in Australia, and for policy, are considered below.

Greenhouse gases and global climate

Greenhouse gases warm the earth by allowing the sun's radiation to penetrate the atmosphere and be absorbed by the planet but restricting the re-radiation of this longer wave radiation back out into the space of the solar systems. The minute quantities of the greenhouse gases mentioned above (CO_2, CH_4 and N_2O) in the earth's atmosphere—that 100-km-thick gaseous layer (blanket) that surrounds our planet—are responsible for the global mean temperature of approximately 15°C. Physicists have calculated that without these greenhouse gases in the atmosphere the global mean temperature would be approximately −15°C with much greater day/night temperature differences, like those of the planet Mars. It is extremely unlikely that life would have evolved on this planet under those climate conditions. These greenhouse gases become completely mixed around the globe each year by global circulation systems. Therefore the release of greenhouse gases is a global not a local problem, because no matter what country releases the gases they will be evenly mixed in the atmosphere above all countries within 12 months of their release.

Carbon dioxide is responsible for 69 per cent of global warming and its concentration in the atmosphere has increased by 35 per cent in the last century to reach the current 383 parts per million (ppm). Analysis of gas bubbles in Antarctic ice cores has revealed that this is the highest CO_2 concentration for at least 650,000 years, well before modern humans appeared on earth. If the current rates of fossil fuel (coal, oil and gas) burning and deforestation continue, we expect the atmospheric CO_2 concentration to reach 450 ppm by 2050 and 600 ppm by 2100.

Methane (CH_4) and nitrous oxide (N_2O) are more potent greenhouse gases than CO_2, but are present in lower concentrations. To compare and calculate the impact of each greenhouse gas on global warming, each is assigned a global warming potential (GWP) in terms of its potency compared to CO_2, which has a GWP of 1.

Methane concentrations in the atmosphere have risen by 140 per cent in the last century to be 1.85 ppm currently. However, methane has a GWP of 21 and therefore accounts for 19 per cent of global warming. The major sources of this increase in methane release to the atmosphere are from coal mining, municipal waste dumps, ruminant farm animals, and rice cultivation. Because the latter sources, particularly ruminant farm animals (enteric methane), are agriculturally related, methane emissions are most important for agricultural

FIGURE 13.2 Increases in carbon dioxide concentration in the earth's atmosphere at Cape Grim (Tasmania) over the past 30 years

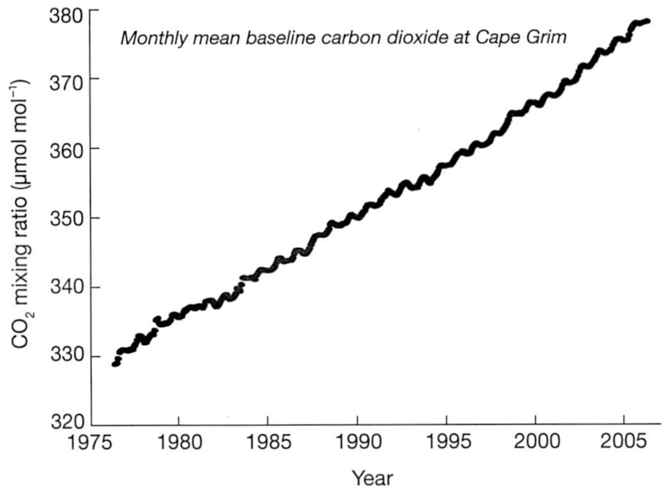

industries. In Australia, methane is 65–70 per cent of agricultural emissions and 60 per cent of national methane emissions.

Nitrous oxide is an even more potent greenhouse gas with a GWP of 310 and contributes 12 per cent of global warming. Atmospheric N_2O concentrations are currently 0.315 ppm and have increased by 20 per cent in the last century, which is largely due to the application of nitrogenous fertilisers in agriculture. Nitrous oxide is released from the soil or animal manure dumps when ammonia is converted to nitrate through a microbiological process. The amount of nitrogen lost as N_2O depends on the soil type and the temperature and moisture conditions. In Australia N_2O is 20 per cent of agricultural emissions and 85 per cent of national N_2O emissions.

In summary, globally CO_2 is the most important greenhouse gas in terms of future global climates and the ever-increasing consumption of fossil fuels. In Australian agriculture, CH_4 and N_2O are the most important greenhouse emissions, which has policy implications because they represent a loss in productivity and a potential cost in any future *emissions trading scheme.*

Climate change and future global climates

The evidence of global climate change and the continued increases in global greenhouse gas emissions raise the important question of how will future climate change impact Australian and global agriculture. The impacts on global agricultural production are important for Australian agriculture because of their impact on global markets for agricultural goods.

Analysis of past climates and climate cycles is of limited assistance in projecting future climates because the 6.5 million people on the earth are changing the planet more rapidly than at any stage in the recorded history of *Homo sapiens*. Therefore, scientists have developed large computer models, called global circulation models (GCMs), of the planet's climate in order to predict what future climates may be locally and globally. One of the key inputs into these GCMs is estimations of how much greenhouse gases will be emitted by future populations and technologies and therefore what will be the future levels of greenhouse gases in the atmosphere. To do this the United Nations has established the IPCC, composed of the world's best meteorologists and climate scientists, to develop a series of scenarios of future development and their influence on future climates. The results of these projections are illustrated in figure 13.3 below.

The different scenarios—A1F1 (high growth, fossil energy intensive), A1B (high growth, balanced energy sources), A1T (high growth, non-fossil energy sources), A2 (low growth, slow technology development), B1 (lower growth, clean energy), B2 (lower growth, more rapid technology development)—indicate that global temperatures are likely to increase from 1.5°C (B1) to 4°C (A1F1) by the end of the twenty-first century (figure 13.3).

FIGURE 13.3 Projections of global greenhouse gas emissions and the resulting global surface warming likely to result in the twenty-first century for each of the IPCC future development and energy generation scenarios

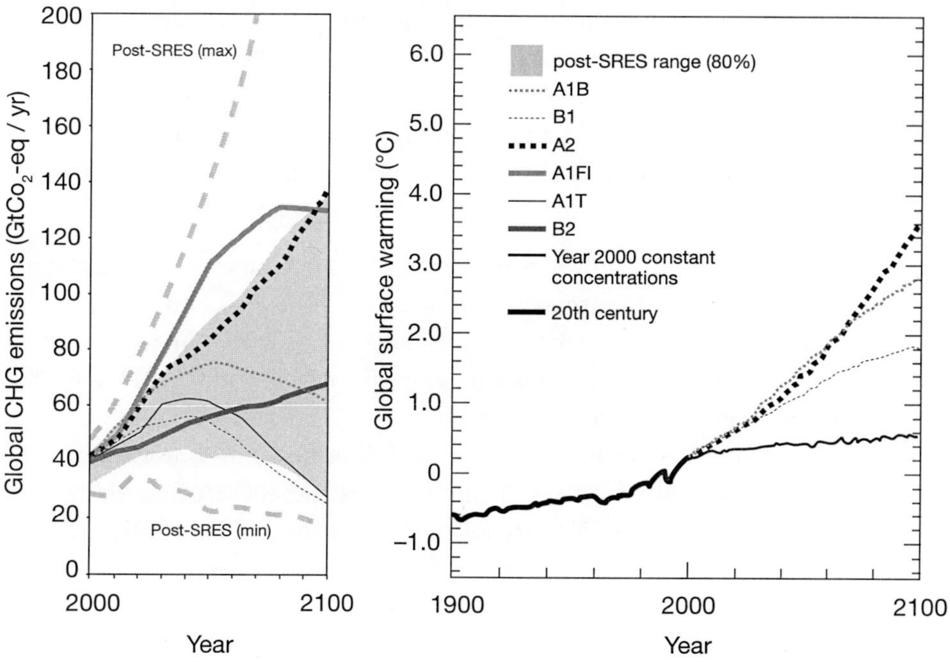

Current global population growth, wealth and fossil energy intensity is tracking a little above the A1F1 scenario, meaning that significant global agreements and action to curb greenhouse gas emissions are needed. As CO_2 remains in the atmosphere for about a hundred years most scientists now believe that a further 1°C of global warming by 2030 is inevitable and 2°C by 2050 is most likely.

Future Australian climates

Given these projected changes to the global climate what will Australian climates be like in the future and how are these likely to have an impact on agriculture? The climate will change in a number of ways of which one or all may impact singly or in combination upon agriculture to produce either positive or negative effects. These include:

- temperature
- rainfall
- available irrigation water from run-off
- increasing carbon dioxide concentrations
- evaporation and plant water use
- frost frequency
- extreme events such as cyclones in tropical areas and heavy rain.

In 2007 the CSIRO and the Australian Bureau of Meteorology released the latest projections for Australian climates for the years 2030, 2050 and 2070 under a number of emission scenarios (CSIRO 2007). These climate scenarios will vary on a regional basis, and increasingly regional climate projections for forward time periods will be available from the CSIRO and the Bureau of Meteorology. As a result of these projections the general patterns of climate change in Australia that we can expect into the future are:

- In 2030 mean temperature increases of about 1°C above 1990 levels are expected, with coastal areas warming less (0.8–0.9°C) and inland areas warming more (1–1.2°C). By 2050 warming of 1.5–2.8°C above 1990 levels is expected for the highest emission scenario (A1F1) and 0.8–1.8°C for the lowest emission scenario (B2).
- There is greater uncertainty with the rainfall projections; however, in 2030 rainfall changes from 0 to –10 per cent are expected in south-eastern Australia, 0 to –15 per cent in south-western Australia and from +5 to –10 per cent in northern Australia. In 2050 rainfall decreases of 7.5 per cent for the A1F1 scenario and 5 per cent for B1 are projected for southern Australia. The projected changes in state average rainfalls are given below for a high emission scenario (table 13.1).

TABLE 13.1 Projected changes (% change relative to 1990) in statewide average rainfall in 2030, 2070 and 2100 (Garnaut 2008)

YEAR	NSW	VIC	QLD	SA	WA	TAS	NT	ACT
2030	−2.5	−3.5	−2.4	−4.2	−4.1	−1.4	−2.5	−2.8
2070	−9.3	−12.9	−8.6	−15.5	−14.9	−5.1	−9	−10.3
2100	−13.7	−19	−12.7	−22.8	−21.9	−7.6	−13.3	−15.2

- Run-off and stream flow are very sensitive to changes in rainfall resulting from climate change and will vary greatly from catchment to catchment. For example in 2030, the 5 per cent average decrease in rainfall in south-eastern Australia could result in a 15–22 per cent decrease in run-off and stream flow. In south-western Australia the 10 per cent decrease in rainfall in the past thirty years has resulted in a 65 per cent decrease in water flow into Perth's storages.
- The increasing temperature will also increase the evaporation rate and therefore the water requirements of plants. In 2030 there could be 2–4 per cent increase in evapo-transpiration rising to 4–6 per cent by 2050 (CSIRO 2007).
- The atmospheric CO_2 concentration is projected to be 450 ppm by 2050 and 600 ppm before the end of this century. In addition to being a greenhouse gas, CO_2 is also an essential plant nutrient that is in short supply. Yield increases of 15–25 per cent with concomitant increases in water use efficiency in temperate (C3) crops such as wheat have been demonstrated in the field at CO_2 levels of 450–500 ppm. Tropical (C4) crops such as maize and sorghum also respond to elevated CO_2 at these levels but not as strongly (10–15 per cent).
- Frost frequency is expected to decrease with global warming but not as much as warming temperatures would indicate. In inland Australia the frequency of frost may decrease by as much as 30–50 per cent by 2050. However, it is important to note that the risk of frost damage to particular crops may not decrease as much because of faster and more advanced crop growth.
- Extreme events including extreme temperatures (>35°C), high winds (cyclone areas) and high-intensity rainfall are all expected to increase in a warming climate. An indication of the changes in the number of hot days above 35°C in each capital city in 2030, 2070 and 2100 is shown in table 13.2.

Australia is a large country with many climatic zones and therefore the adaptation of agriculture to climate change may involve agricultural industries gradually moving between these zones, and practices and genotypes being imported from one to the other. This could well involve a number of considerations, which are discussed below.

TABLE 13.2 Projected increases in days above 35°C for all Australian capital cities (Garnaut 2008)

	CURRENT	2030	2070	2100
Melbourne	9	12	21	27
Sydney	3.3	4.4	9	14
Brisbane	0.9	1.7	8	21
Adelaide	17	22	34	44
Perth	27	35	56	72
Canberra	5	8	21	32
Darwin	9	36	221	312
Hobart	1.4	1.7	2.5	3.4

Adaptation of Australian agriculture to climate change

The development and progress of Australian agriculture over the past 150 years has been one of adaptation to a highly variable climate, emerging disease threats and constantly changing global markets. Throughout this period Australia's agricultural industries have collectively maintained productivity gains in the order of 2 per cent per year through the application of science and economics to the development of innovative farming systems. Climate change through the projected changes in temperature, rainfall, CO_2 and so on will pose a significant challenge to agricultural industries and the research networks that service them. However, the early identification of potential threats to particular industries in particular regions will allow the application of science and economics to the identification of appropriate and profitable adaptation strategies. The key to these responses will be early and decisive action in relation to our international trade competitors.

Preliminary studies by the Australian Bureau of Agricultural and Resource Economics (ABARE) (Gunasekera et al. 2007) indicate that Australian agriculture may be disadvantaged in relation to our global competitors through a limited capacity to adapt to climate change (figure 13.4).

On the basis of this ABARE study four of Australia's major agricultural export industries—wheat, beef, dairy and sugar—are likely in the year 2050 to be affected by climate change more than those industries of our competitors, without the development of significant adaptation and mitigation strategies. To avoid these detrimental effects on the international competitiveness of Australian agriculture effective mitigation and adaptation measures must be devised and implemented. When significant adaptation and mitigation strategies and the potential ameliorating effects of increasing CO_2 are considered, a different picture may emerge for Australian agriculture, as the example of the potential of the wheat industry in table 13.3 indicates.

FIGURE 13.4 The potential impacts of climate change on production of selected agricultural industries in major exporting countries at 2050, without mitigation or adaptation—expressed as % change relative to the reference case

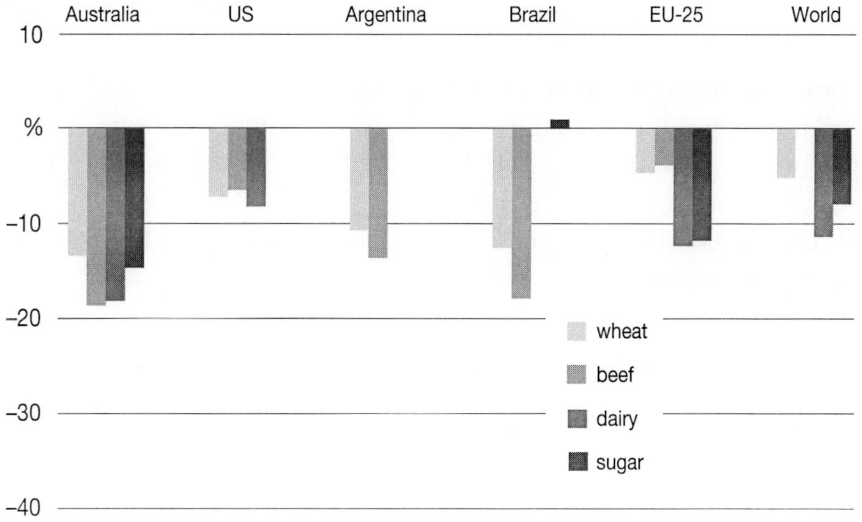

Source: Gunasekera et al. 2007

TABLE 13.3 Adaptation of Australian wheat to climate change under three climate cases—% cumulative yield change from 1990 (Garnaut 2008)

WHEAT REGION	NO STABILISATION		STABILISATION AT		STABILISATION AT	
	650 ppm CO_2 IN 2100		550 ppm CO_2 IN 2100		450 ppm CO_2 IN 2100	
	2030	2100	2030	2100	2030	2100
Dalby, Qld	8.2	−18.5	4.8	−1	1.6	−3.7
Emerald, Qld	7.2	−10.1	4.4	0	1.8	−2.5
Coolamon, NSW	11.6	1.9	9.9	12.3	8.2	7.4
Dubbo, NSW	8.1	−5.9	6.1	6.7	4	2.3
Geraldton, WA	12.5	22.4	9.7	5.6	6.9	2.6
Birchip, Vic	14.8	−24.1	10.7	1.5	6.8	−0.3
Katanning, WA	15.6	16.8	14.8	18.9	13.9	14.6
Minnipa, SA	0.8	−23.9	−3.4	−15.3	−7.4	−15.7
Moree, NSW	20.6	10.9	17.7	14.1	14.8	10.8
Wongan Hills, WA	16.1	−21.8	13	5.5	10	4.4

The examples given above illustrate the ongoing need for further research and analyses in developing appropriate climate change adaptation strategies for particular industries in specific regions. In developing these strategies the key considerations will be:

- development of good regional climate change scenarios for the future
- likely impacts on the water resources for the region, particularly in irrigated areas
- availability of more heat-adapted varieties
- potential opportunities to introduce new more productive crops
- potential ameliorating capacity of increased CO_2 on yields
- risks of damaging extreme events
- likely impact of climate change on global supply and demand for a particular product or commodity.

Mitigation of greenhouse gases in agriculture

In national and international terms, agriculture is a relatively large emitter of greenhouse gases per unit value of product produced. For example, in Australia, agriculture contributes about 3 per cent of the national income while emitting approximately 17 per cent of the nation's greenhouse gases. If the indirect emissions from the fertiliser and chemical manufacture and transport fuels are also considered, agriculture is responsible for more than 25 per cent of the national emissions. The greenhouse emissions from a typical agricultural enterprise are given in table 13.4. The impact of methane emissions from livestock (sheep) is immediately apparent, contributing almost 60 per cent of all farm emissions. The total farm emissions from the mixed farm were almost four times that of a similar sized cropping farm. The fuel and fertilisers that depend on fossil fuel comprise only 21 per cent of greenhouse emissions on the wheat–sheep farm but about 58 per cent on the cropping farm.

As the globe seeks to avoid serious climate change (>2°C) by restricting further increases in greenhouse gas emissions through international agreements such as the Kyoto Protocol, agriculture will have to consider how to reduce (mitigate) all or part of these emissions. In national terms the proposed introduction of an emissions trading scheme requiring all emitters to purchase emission permits is a considerable threat to agricultural industries, assuming agriculture is included in the emissions trading scheme.

Agriculture is not currently included in initially proposed greenhouse gas caps and emissions trading schemes. As agriculture is a major contributor to greenhouse gases, this situation will change. Incorporating agriculture into greenhouse emissions trading will be difficult, because it is difficult and costly to measure and monitor agricultural outputs of greenhouse gases. Regardless, the costs of greenhouse emissions in other parts of the economy, and

TABLE 13.4 The greenhouse gas emissions profile of typical mixed farming and cropping businesses in the wheatbelt (Keogh 2007)

EMISSION SOURCE	GREENHOUSE EMISSIONS		% OF TOTAL
	GAS	TONNES $CO_{2\text{-}e}$	
SHEEP–WHEAT FARM: 2200 HECTARES			
• Electricity and fuel	CO_2	123	5
• Livestock: 4000 sheep	CH_4	1562	58
• Nitrogen fertiliser: 70 tonnes	N_2O	426	16
• Soils: nitrogen	N_2O	409	15
• Animal wastes: nitrogen	N_2O	170	6
TOTAL		2689	100
WHEAT–CANOLA FARM: 1700 HECTARES			
• Electricity and fuel	CO_2	162	25
• Nitrogen fertiliser	N_2O	246	37
• Soils: nitrogen	N_2O	207	31
• Burning	CO_2	45	7
TOTAL		660	100

steps taken to reduce these emissions, will incur costs that will be borne by consumers and producers throughout the economy. Some of these costs will be borne by farmers.

In the scenario that agriculture is not included in the emissions trading scheme it is likely that direct (fuel and power) and indirect (fertiliser) energy-related inputs will be subject to price rises as a result of having to purchase permits. In this case the cropping farm will experience greater increases in its input costs. If agriculture were included in the scheme and had to purchase permits for all its emissions the mixed farm would be more greatly affected. The potential change in these costs at a permit price of \$40/tonne $CO_{2\text{-}e}$ are illustrated in figure 13.5.

In this analysis it is apparent that production costs of a cropping and livestock enterprise would increase by 4.5 per cent and 2.5 per cent respectively if fossil fuel energy–related inputs only were included in a scheme. However, if agriculture is included, and continues to operate as it currently does, the production costs in the cropping and livestock industries rise by 6 per cent and

18 per cent respectively. These scenarios dramatically illustrate the vulnerability of agriculture if included in an emissions trading scheme in the absence of any compensating arrangements and without any farm adaptation.

FIGURE 13.5 The impacts of purchasing carbon emissions permits on the cost of agricultural production for the cropping and livestock industries for either agriculture's inclusion or exclusion in an emissions trading scheme that valued carbon at \$40/tonne CO_2 (no adaptation).

Source: Gunasekera et al. 2007

The future for agriculture

In this chapter we have learnt that climate change is an important issue for agricultural managers to be aware of and to plan for. The climate has changed and is expected to change more rapidly during this century, which will necessitate a range of adaptive responses from farmers. Agricultural industries and regions can expect temperature increases of 1°C by 2030 and a further 1°C by 2050 with a strong possibility of decreased rainfall in southern Australia.

As a result of global concerns about climate change and its potential impacts on food production and human well-being it is likely that global action will be taken to reduce emissions and avoid dangerous climate change. Agriculture is particularly vulnerable to rising production costs if emission trading schemes are introduced to reduce emissions because of its relatively high greenhouse gas emissions per unit of production.

Conclusion

There are a number of themes running through this book, which seeks to give an introduction to agriculture in Australia. In its representation of the various themes, agriculture is characterised in the following ways:

- Agriculture is one of Australia's few world-class activities.
- Being involved in agriculture is exciting, intellectually challenging and highly rewarding.

- People working in farming and agriculture-related businesses have a love for agriculture.
- Farming in Australia is not an easy way to make a living, but those who are good at it and whose businesses are medium-sized or larger can make returns to capital that are as good as any other business in the economy.
- There is a lot of risk in agricultural businesses, which farmers have to manage well to make a good living. The risks are mainly associated with the weather affecting the size of crop harvests and the supply of feed for livestock, and the prices received in the markets for farm commodities.
- Farmers are highly skilled at adapting what they do to the changes that occur constantly in the economy and the natural environment.
- Good science has a huge role to play, discovering and developing new and more efficient ways to grow farm products.
- Ratios of average input to output of single inputs to production, such as milk per cow or wheat per hectare or energy in and out, or water in or out, are called technical efficiency. This measure is not the correct criterion to use to decide what to do in a farm system. The ratio of benefits to costs, or the income versus cost of all inputs, known as economic efficiency, is the criterion to decide how to operate a farm system.
- The whole farm approach is the only way to analyse how well a farm business is performing and the only way to evaluate whether a change to a farm system is worth doing or not.
- For management purposes it is more useful to solve the whole of the problem in an approximate way than solve part of a problem extremely well.
- The key principles that determine how well a farm business performs are the principle of comparative advantage, the principle of diminishing marginal returns, the principle of increasing risk and the probability principle.
- What happens where and how and when in agriculture in Australia is determined by the natural conditions, especially rainfall, and equally, by whether or not the farming activity is profitable.
- Farming in Australia has always been mostly a hot and dry activity, and will continue to be so.

Ultimately, it is farmers and people who invest in and run and work in businesses that transform farm product into consumer goods and distribute it to markets for consumers to buy who will, together, ensure that the growing population of the world is fed adequately with the quantity and quality of food they require. A central belief in this book is that a mix of good science and good economics applied to agriculture at farm level and beyond the farm gate is critical to agriculture being able to supply the food requirements for a rapidly growing world population.

Investment in agricultural scientific research has declined in recent decades and such investment is one of the keys to increasing world food supply. The other key to meeting world food needs is having economic policies that enable farmers around the world to produce efficiently and trade freely, without the distortionary effects of trade barriers and domestic interference with production and prices such as subsidies on production and taxes on exports that discourage good farming. The solutions to growing world food demand are both scientific and economic.

A major motivation for writing this book is serious concern about the growing disconnection between people in increasingly urbanised societies that have rapidly growing non-farm sectors of the economy around the world and people working in agricultural production and marketing. This disconnection between agriculture and non-agriculture sectors of the population is leading to increasing, widespread ignorance of the challenges and rewards of working in agriculture, both within and beyond the farm gate.

The ultimate effect of this disconnection between people who have and those who do not have an understanding of agricultural science and economics is that agricultural production and marketing as a business and research activity is struggling in the competition to attract the most essential resource of all—the best and the brightest young minds. This is a serious issue. The quantity and quality of natural resources is not the limiting factor that will constrain the capacity of the agricultural sector to feed the world. The limiting factor to world agricultural production is the quality of the knowledge of human resources working in the agricultural sector.

REVIEW QUESTIONS

1. What change can we expect in the climates of Australia's agricultural regions in the next twenty and forty years respectively?
2. Discuss the likely impacts of climate change on the Australian wheat industry in the medium and longer term and how the industry may adapt to it.

FURTHER READING

Australian Bureau of Meteorology, *The Greenhouse Effect and Climate Change*, Melbourne, 2006.

CSIRO, *Climate Change in Australia*, Technical Report, CSIRO, Melbourne, 2007: www.climatechangeinaustralia.gov.au.

Garnaut, R,. *The Garnaut Climate Change Review*, Cambridge University Press, 2008.

Gunasekera, D., Kim,Y., Tulloh, C. and Ford, M., Climate Change—Impacts on Australian Agriculture, *Australian Commodities* 14:4, ABARE, Canberra, 2007.

Intergovernmental Panel on Climate Change, *Climate Change 2007: Impacts, Adaptation and Vulnerability*, WGII contribution to the IPCC Fourth Assessment Report, 2007: www.ipcc.ch/.

Intergovernmental Panel on Climate Change, *Climate Change 2007: Mitigation*, WGIII contribution to the IPCC Fourth Assessment Report, 2007: www.ipcc.ch/.

Keogh, M., *The New Challenge for Australian Agriculture: How do you Muster a Paddock of Carbon?* Australian Farm Institute, Surry Hills, 2007.

Pittock, B., *Climate Change—An Australian Guide to the Science and Potential Impacts*, Australian Greenhouse Office, Canberra, 2003.

WEBSITES

Australian Bureau of Meteorology: www.bom.gov.au

Australian Farm Institute: www.farminstitute.org.au

Australian Government: www.climatechange.gov.au

Australian Projections: www.climatechangeinaustralia.gov.au

CSIRO: www.csiro.au/org/ClimateAdaptationFlagship

Garnaut Climate Change Review: www.garnautreview.org.au/domino/Web_Notes/Garnaut/garnautweb.nsf

Intergovernmental Panel on Climate Change: www.ipcc.ch

National Oceanographic and Atmospheric Administration (NOAA): www.noaa.gov/climate

Glossary

activity
a particular method of producing a commodity. More specific term than 'enterprise' e.g. spring wheat, winter fattened steers.

activity gross income
the total value of the output of a farm activity whether the output is sold or not.

activity gross margin
activity gross income minus the variable costs of that activity.

activity gross margin planning
a procedure whereby activities are selected sequentially on the basis of the highest gross margin from a unit of only one key constraint, usually land.

agribusiness
business that is closely related to agricultural production activities.

amortised loan
a loan that is repaid in equal instalments of principal and interest, with the interest and principal components of the repayment instalment varying as the loan reduces.

annuity
a sum of money received or used every year that is equivalent to a larger sum at the present time or at a future time, e.g. the equal annual sum that repays the interest and principal on an amortised loan. An annual sum over a number of years that is equivalent to the net present value of an investment project that runs over the same number of years.

break-even analysis
testing key elements of a budget of a proposed change to a system to determine the level at which key elements of the proposal make the overall result just equal to the result from an alternative action.

budget
a detailed statement of a future plan of action detailing the expected costs and benefits.

budget control
the process of comparing the actual performance of an aspect of farm production against the performance that was expected when the budget was drawn up.

business health
the state of and prospects for profit, financial viability and growth of a business.

capital
items that contribute to production over at least a medium time period, such as for more than a year (for example, tractor, land, lime fertiliser, structures and equipment).

capital gains
increase in the value of capital items due to a rise in their market value.

capital investment
funds used to acquire assets such as equipment, land or stock, or used on improvements that have a life of more than one year and add to the productive capacity of the farm.

cash flow
the movement of funds in and out of the hands of an enterprise or individual farmer.

cash flow budget
a budget of the expected cash in (receipts) and cash out (payments) associated with a particular farm plan.

cast for age (CFA)
a reject old animal that is past its economic life for particular conditions.

comparative analysis
comparison of the performance of a particular farm with some 'standard' level of performance. (Usually the 'standard' is the average performance of a group of broadly similar farmers.) The term 'benchmarking' is used to describe a similar approach.

core discipline
an integrating body of knowledge. Integrates knowledge from many disciplines into comprehensive understanding, analysis and explanation. The core discipline of economics makes it possible to understand, analyse and explain the operation of a whole farm system.

debt servicing capacity
annual whole farm net cash flow available to meet interest and loan repayments.

decision analysis
a procedure for rigorously and methodically weighing up the expected benefits and costs of a possible action. A way of ensuring that decision makers make decisions that are consistent with their personal beliefs about the risks they face and their personal preferences for possible consequences from the decision.

demand
the amount of a product or service that consumers wish to buy at a range of prices.

depreciation
the loss in value of capital items as they get used and become older.

diminishing marginal returns
the phenomenon that increases in variable inputs added to fixed inputs in a production process results in smaller and smaller increases in total output. The principle of diminishing marginal returns indicates that variable input should be added to the production process so long as the extra return exceeds the extra cost, and the maximum total profit is at the point where extra return equals extra cost.

discounting
the process of adjusting the value of a benefit or cost to be received in the future to their equivalent value at the present time.

discounting factor
the adjustment factor used to adjust future values to present values, given by the formula 1 ÷ (1 + r)^n, where 'r' is discount rate and 'n' is the number of the year in the future in which the benefit or cost occurs.

economic efficiency
the amount of profit made by a business. Measured as the profit expressed a percentage of the total capital invested in the business.

elasticity of demand
the change in the quantity of a product or service people will buy when the price changes. Expressed as a percentage change in quantity divided by percentage change in price.

enterprise
the production of a particular commodity or group of related commodities. A general term (for example, wheat or beef enterprise).

equity
the value of assets minus liabilities. Also known as net worth. What the business owes the owners.

equity percentage
farm equity capital as a percentage of total farm capital—that is, (assets minus liabilities/assets) × 100/1.

farm benefit cost analysis
the budgeting process of evaluating the benefits and the costs and the net benefits of an investment to change a farm system.

fixed capital
land, buildings, bores, irrigation equipment etc. that cannot easily be moved.

fixed costs
costs that must be met and are not affected by the amount of output produced in a year. Also called overhead costs. They are unavoidable costs in the short to medium term.

gearing
the ratio of debt to equity. Has implications for debt servicing ability and rate of growth of equity. (Also called leverage.)

gross margin
gross income minus variable costs. Can be whole farm gross margin, as in a whole farm budget, or activity gross margin.

growth
increase in net worth (wealth) over time. Measured as change in equity,

or net farm income minus tax and consumption above operator's allowance.

inflation
an increase in the supply of money in relation to the supply of goods and services available and, in consequence, a decline in the purchasing power or value of currency.

innovation
changing how a farm system operates.

interest
the annual sum a lender charges someone who borrows funds from them. Expressed as a percentage of the sum borrowed—for example, 10 per cent interest p.a. on $100,000 borrowed.

interest-only loan
a loan where the borrowed capital is not intended to be repaid on a regular and gradual basis over the life of the loan and instead annual interest is paid on the full amount of the borrowed capital for the life of the loan.

liquidity
cash or near-cash reserves. Relates to the ability of a business to service debt.

marginal
economists' word for extra or added. The principle of marginality refers to the profit-maximising level of operation where the marginal revenue from production equals the marginal cost of production.

marginal cost
the extra cost added to total cost from using an extra unit of a variable input; or, the extra cost incurred in growing or selling an additional unit of product.

marginal revenue
the extra net income obtained from selling one additional unit of product.

marketing
all decisions made by an entity that determine the characteristics of the output (the marketing mix) presented to potential customers.

marketing system
the entire system composed of entities contributing to the production of a category of product; also, in agricultural product contexts, called 'agribusiness system'.

monopoly
there is only one seller of a product or service.

net cash flow
the difference between the money received and the money spent in any one period (week, month, year).

net farm income
operating profit minus interest. Also called net profit. Is the return to the owner's capital.

net present value
the difference between the present value of all benefits and present value of all costs of an investment, with the present values of benefits and costs calculated using a particular discount rate.

net worth
the value of total assets minus the value of total liabilities (equity).

operating costs
variable costs plus overhead costs.

operating profit
gross income minus variable and fixed (overhead) costs. Is the return to all the capital invested.

opportunity cost
the opportunity cost of a farm management decision is the amount of net benefit that is given up by choosing one alternative action rather than some other action.

overhead (fixed) costs
costs that do not vary as the level of production or mixture of activities changes. Are unavoidable costs in the short to medium term.

parameter
any factor that has an important effect on profit (yield, price, hectarage, direct cost).

partial budget
a budget drawn up to estimate the effect on profit of a proposed change affecting only part of the farm. Used to estimate the extra return on extra capital invested.

principal
the amount of capital borrowed when a loan is taken out. Principal repayments are the amounts of capital repaid to settle a debt.

production function
the relationship between the level of inputs and the level of output for some production process. Also called a response function.

real terms
dollar values or interest rates that have no inflation component.

return on total assets
operating profit expressed as a percentage of the value of total farm assets.

return to total capital
the annual operating profit expressed as a percentage of the total capital invested in the business over the year. (Total capital can be capital at start, end or, better, an average of start and end capital value.) Percentage return to total capital is the measure of economic efficiency.

risk
a situation with uncertain outcomes, but a case where some probabilities can be formed about the outcomes. This is in contrast to uncertainty, where no probabilities can be formed about uncertain events happening.

solvent
assets exceed debts.

stock equivalents
units used in livestock feed budgeting whereby the energy needs of different categories of livestock are expressed in terms of a standard livestock unit, which is a 4.5 kg wether, called a dry sheep equivalent.

supply
the amounts of a product or service that will be offered for sale at a range of prices.

technical efficiency
a ratio of the quantity of physical output to quantity of physical input. Does not indicate profitability or economic efficiency.

term loan
a loan that is to be repaid in equal annual instalments of principal, with interest charged on the reducing outstanding balance of the loan.

variable costs (also direct costs)
costs that change directly according to the amount of output of the activity, e.g. fuel, seed.

whole farm approach
the farm management economic method. Understanding and analysing the farm system—the human, technical, economic, financial, risk, institutional elements—as a whole system.

Index